The Open University

A Second Level Course

INTRODUCTION TO ENGINEERING MECHANICS

Unit 1
Introduction and Mathematics

Prepared by the Course Team

THE OPEN UNIVERSITY PRESS

The Introduction to Engineering Mechanics Course Team

Authors

J. K. Cannell	(Engineering mechanics) Chairman
O. R. Fendrich	(Engineering mechanics)
G. S. Holister	(Engineering science)
P. J. Lucas	(Engineering mechanics)
V. Marples	(University of Warwick)
P. Minton	(Imperial College, University of London)
R. K. Pefley	(University of Santa Clara, California)
C. N. Reid	(Materials)
B. O. Shorthouse	(Staff Tutor)

Other members

A. J. Crilly	(BBC)
R. P. Dobson	(Course Assistant)
R. D. Harrison	(Educational Technology)
A. R. Sollars	(Staff Tutor)
S. J. Stickland	(Editor)
V. Woodhouse	(Technician)
P. R. V. Youngman	(Scientific Officer)

The Open University Press
Walton Hall, Milton Keynes

First published 1975

Designed by the Media Development Group of the Open University.

Martin Cadbury, a specialized division of
Santype International, Worcester and London

ISBN 0 335 02850 0

This text forms part of an Open University course. The complete list of units in the course appears at the end of this text.

For general availability of supporting material referred to in this text, please write to the Director of Marketing, The Open University, PO Box 81, Milton Keynes, MK7 6AT.

Further information on Open University courses may be obtained from the Admissions Office, The Open University, PO Box 48, Milton Keynes, MK7 6AB.

1.1

CONTENTS

AIMS

The text of this unit has been written in two parts.

The first part contains a few short stories and a little history. It is meant to give you – who are perhaps a newcomer to the field of engineering – some idea of what engineering mechanics is about. It has no definite aims or objectives but it is intended that, at the least, you may enjoy reading it and hence look forward to the rest of the course. You may learn from it as well.

The other larger part is intended for revision of mathematics. It should cover the extra mathematics needed to enable anyone with school 'O' level mathematics ability (perhaps learnt a fair time ago) to enter the course successfully. Most of it will be revision to many, but the aim of this unit is to provide a common starting level, from which, mathematically, we can jump off into succeeding units.

OBJECTIVES

There are no special objectives in the first part of this unit. The following objectives refer to the second part on mathematics for mechanics. After reading this part you should be able to:

1   Use and understand the symbols listed in Section 2.1.

2   Use brackets correctly.

3   Express repeated factors in terms of the factor raised to a power, whether the latter is positive, negative or fractional.

4   Express the root of a number in terms of a fractional power.

5   Use the tables of logarithms to carry out multiplication, division, raising to a power and any combination of these operations.

6   Use the tables to find the sine, cosine and tangent of any angle between $0°$ and $360°$.

7   Sketch the graph of the general equation $y = mx + c$ and understand what each term means.

8   Sketch the graph of the general equation $y = ax^2 + bx + c$ for various values of the constants.

9   Appreciate that the slope of a parabola changes at a constant rate.

10 Solve simultaneous non-linear equations where the result depends on the solution to a quadratic equation.

11 Know what the slope of a curve at a point means.

12 Calculate the slope at a point on a curve having a known equation.

13 Differentiate a function of a variable with respect to the variable to obtain the first or second differential of the function.

14 Relate the instantaneous velocity and acceleration of a point moving according to some function of time, to the corresponding first and second differentials of the function with respect to time.

15 Differentiate a product, a quotient and a function of a function.

16 Relate the integral of a function to the area under the graph of that function.

17 Relate the constant of integration to the initial conditions.

18 Evaluate definite integrals.

19 Differentiate and integrate simple exponential functions.

STUDY GUIDE

No matter how extensive or limited your background in maths, make sure you can handle the maths in this unit as specified in the objectives. Be sure to test yourself critically using the SAQs in this text and the problems on your tutorial example sheets.

If maths horrifies you, read the introduction to Section 2 twice and be reassured. Even if your mathematical ability is low at this moment, after you have worked through Part 2 of this unit you should be well able to progress through the rest of the course with enjoyment.

In the unlikely event that most of the maths in this unit is new to you, it may be more work for you than we anticipated. In this case, return later to finish this unit: do not get behind with subsequent ones. The density of maths in this unit is *not* typical of that in the remainder of the course.

There are no discs, television programmes or experiments associated with this unit.

Part I

INTRODUCTION

1.1 Man against nature

Railways had been prevented from penetrating directly into the east of Scotland by the two mile width of uncomfortable water in the Firth of Tay. Twice in the journey to Dundee or north-east Scotland from Edinburgh or the north of England, passengers had to get out of their trains and cross the water in small ferries.

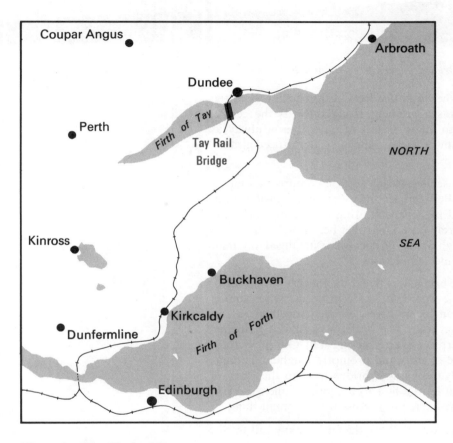

Figure 1   The Firth of Tay

In 1871 a contract for the construction of the first bridge to span the Tay (across about two and a half miles) was let to De Bergh and Company. After three years it was transferred to Hopkins, Gilkes and Company of Middlesborough, who took over most of the local site plant owned by their predecessors. This included a foundry at Wormit, on the south bank of the estuary, where most of the ironwork for the bridge was cast.

The designer of the bridge throughout was the engineer Thomas Bouch, whose original design was for a lattice girder bridge carrying a single rail track over its 200 foot girder spans. The bridge was to be supported on what were allegedly rock foundations in the Tay by brick piers. The bedrock was not solid, however, so Bouch redesigned his bridge with cast-iron columns mounted on a circular masonry base 27 feet in diameter, instead of all-brick piers. This new design had the central main girder spans increased to 245 feet, except for two of 227 feet. The total number of 85 columns were cast in sections, bolted, braced with wrought iron and partly filled with concrete.

Figure 2   The Tay Railway Bridge, viewed from the north

The track was laid on top of the 84 lattice iron girders for most of the length of the bridge. There was a gradient rising towards the centre, where 13 girders were raised to allow 88 feet of navigable headway at high water. Here the track ran lower, *inside* these intricate, latticed girders, near to their lower flanges.

The whole bridge was built in sections, each section comprising four or five of the great girder spans. Between sections was an expansion joint, and each section was attached by a fixed bearing to one pier and was simply supported (that is it merely rested on roller bearings) on the other four or five piers. The 13 'high span' girders were arranged on their 14 iron piers in three inter-connected sections.

When completed, this immense bridge was by far the longest in the world. It was stringently tested by the Board of Trade inspector who, among other things, ordered six 73 ton locomotives, coupled together, to run over the bridge at 40 miles per hour. The bridge was officially opened in 1878 and acclaimed a masterpiece of engineering skill. Queen Victoria crossed it in style with a salute of guns and promptly knighted Thomas Bouch, who was by then busily designing a bridge to span the Firth of Forth, north of Edinburgh. To be the first to accept a challenge on so ambitious a scale and have any degree of success was no small thing.

For eighteen months the bridge proudly bore the fairly heavy rail traffic. Soon after its opening, however, its maintenance engineer had observed that many of the wind-bracing ties, supposed to be in a permanent state of tension to stiffen the girders and columns, were loose and rattling, but it was claimed that this matter was rectified.

Then on Sunday December 28th 1879 a great storm arose. The wind, carrying scudding rain squalls, became so strong that a local signalman on the approaches to the bridge had to crawl on hands and knees to make his way to the shelter of his cabin. At seven o'clock in the evening the last passenger train, carrying mail, arrived at St Fort station on the southern approach to the Tay Bridge. By this time the fury of the storm had increased still more. The train, powered by a (then) modern 4-4-0 express locomotive No. 224, was permitted to enter the last section on its journey to Dundee.

The 'train entering section' signal was telegraphed to the north shore signalman, and a few people saw the tail lights of the train disappear among storm and girders into the roaring night. Inexplicably, for a while, they saw a trail of sparks flying from its wheels. Then a particularly violent gust hit the cabin, there was a flash over the bridge, and nothing more was to be seen.

The St Fort signalman found that his instruments and the Dundee telegraph line were dead. He tried to follow the train out on the bridge, but it was impossible to do so in that wind. The water supply, piped over the bridge from Dundee, also failed.

A little later the north-bank signalman became worried that the expected train had not appeared. Two or three accidental observers at different places near the shores of the Tay thought they saw glimpses of 'sparks or streams of fire' descending from the bridge. At Dundee station there was much consternation, and a station foreman and an attendant tried to make their way along the bridge to investigate.

With no communications across the Firth some optimism remained, until later again in the evening some train wreckage and mail bags were found washed ashore at Broughty Ferry, a few miles towards the mouth of the estuary. Finally at ten o'clock the Tay ferry steamer somehow managed to struggle across to Dundee to establish the certainty of catastrophe.

The other reports slowly came in. By the light of the moon through occasional gaps in the flying clouds, a few people saw that that 13 high spans of the mighty Tay Bridge, those three special sections above the navigable river, were all missing.

When the storm subsided, the train was found in the river, lying within the wrecked central girders of the high spans. It had not plunged off the bridge through a gap, but its very presence on the latticed high spans had so increased the lateral wind loads on the girders that they, with the train, had literally been blown off their roller bearing supports on the iron piers.

Figure 3   The Tay Bridge disaster: steam launches and a divers' barge employed in the search (Illustrated London News, 10th January 1880)

Apart from the last two coaches the train was not severely damaged, as it had been protected by the girders. However, the passengers and crew had obviously died almost instantly. In fact, 'loco No. 224' was repaired and ran again. For many years no driver was prepared to take her over the rebuilt Tay Bridge, but on the twenty-ninth anniversary of the disaster, Sunday December 28th 1908, she worked the same night mail train to Dundee. She continued work until 1919.*

* *Rolt, R. T. C. (1960) Red for danger, Pan Books Ltd.*

The worst blunders of engineers are no more disastrous than those made in professions such as medicine, law, politics, management and urban planning. Unfortunately, they are better defined and more tangible. Engineers were, and are, often pilloried mercilessly for their mistakes, and Bouch was no exception. His theories for the disaster, based on a derailment of the last two coaches, were utterly rejected. The design and construction of the piers were criticized severely; so were the design and maintenance of the bracing ties and the original workmanship. Clearly, the bridge had not been capable of withstanding the prevailing wind pressures, either in its design or in its completed form.

It seems most likely that one contribution to failure was that the weight of the train on one span of an integral section tended to lift the simply supported end of the next. The added force due to the high wind pressure acting laterally on the train as well as on the lattice girder, was then sufficient for the wind to blow the girder sideways off its supports. Certainly in those days there were no figures for wind load, other than some published a whole century before by Smeaton (the famous lighthouse builder). These figures would be considered dangerously low today.

Many bridges had fallen before, others have fallen since. It was known all too well, even then, from the failures of early suspension bridges that such structures could be aerodynamically unstable – unstable once they were set oscillating in a wind. The multi-span beam bridge over the Tay was aerostatically unstable – unstable in a wind even without being set oscillating.

Bouch died a broken man, not long after the disaster. He had by then prepared a design for an even greater 'master-piece' to span the Firth of Forth: a suspension bridge, this time with two 1600 foot (525 metre) spans which he designed to withstand a wind force of 10 pounds per square foot (47.9 newtons per square metre).

Needless to say the project was cancelled. In its place, the design by Sir John Fowler and Benjamin Baker for the cantilever colossus that still strides the Firth was accepted. This design, after a series of tests, was arranged to withstand a wind force of 56 pounds per square foot – five and a half times that previously contemplated!

Figure 4 The Forth Railway Bridge as it is today

We know today that strength of this order is unnecessary. No wind on earth ever blew to demand this resistance. And this is one reason why no bridge of such immense proportions as the Forth Bridge is ever likely to be built again anywhere in the world. Even today, its dynamic stresses are very low. It carries the heaviest and fastest of trains without a murmur, and is as sound as the day it was built more than eighty years ago.

The Forth Bridge began an era of giant cantilever bridges to suit the needs of the day. Suspension bridges were known to be ill-suited to the shock loads of heavy locomotives; the many piers of beam bridges obstructed the navigable waterways that large bridges had to span. In the case of the Forth Bridge, there was one submerged island in the centre of the Forth upon which a pier could be built, thus demanding a mighty span either side to permit passage of shipping. Two centre spans of 1700 feet (558 m) were to remain the greatest for nearly thirty years. Among cantilever bridges today they are only exceeded by the single 1800 foot (591 m) span of the Quebec Bridge in Canada.

The piers of the Forth Bridge were sunk using the then well tried method of pneumatic caissons. By that time enough was known of the dreaded 'caisson disease' for the men to be able to avoid its worst effects, and in fact no lives at all were lost in the sinking of the foundations. The superstructure was of steel – the first long-span railway bridge to be built of this new material. Fifty-one thousand tons of steel and six and a half million rivets went into the monster, and the steel-work was erected outwards from both sides of each main pier so that the weights on each side were kept in balance.

The ends of the cantilever arms were joined by a suspended span. The final joining of the parts had to be carried out when the temperature

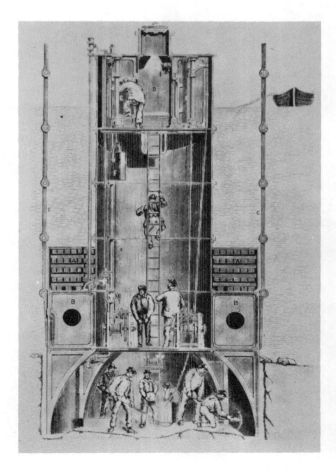

Figure 5   A pneumatic caisson

Figure 6   The steel-work construction of one of the main piers of the Forth Bridge

was exactly right. Steel has a high coefficient of expansion – it expands considerably as it gets warm and contracts as it cools. The gaps between railway lines at the fish-plates are supposed to allow for this change. How much more, then, must such changes matter in something as long as the Forth Bridge.

Figure 7   Disaster on the St Lawrence: the wreck of the Quebec bridge

In an earlier American bridge, the three-arched steel St Louis Bridge, forty-five tons of ice had been packed around the members when the rising air temperature threatened to prevent the last bolts from being rammed home. The Forth Bridge was designed to be completed at a temperature of 60°F (15.6°C). But the weather proved obstinately cold, so wood shavings and oily waste were set alight on the steel to expand it the last few millimetres.

The Scottish mammoth was finished in 1890, after seven years of toil. It remains, perhaps for all time, the King of Cantilevers. Its construction had proceeded smoothly and steadily – so differently from its one possible competitor, the Quebec Bridge over the St Lawrence River in Canada. This – the only longer cantilever in the world – was entirely redesigned after initially collapsing, took twice as long to complete (from 1904 to 1918), and was one long succession of tragic failures.

In one accident 9000 tons of steel collapsed into the river killing 75 men. In another, owing to a defect, its 5000 ton suspended span crashed and 13 more men died. Like most great adventures, the building of the Forth Bridge killed too: up to 4500 men worked on it and 57 lost their lives. However, in the confident words of the official reports of the time, these deaths were mainly due to the men's own carelessness.

It is a compelling story. Clearly, many lessons were learnt during those years of designing for the unknown. Of course, it is always easy to be wise after the event. The reasons for the disasters usually became obvious when investigations were carried out later, and nowadays any competent engineer would pick out the mistakes made in those days as being fairly elementary.

You may also think they are, and may wonder why they did not seem so simple to the designers in those days. One reason is perhaps that there was not so much experience of bridge failures then as there is now. The main reason however is simply that not enough was known about the properties of materials and how various structures behaved under various conditions.

The ignorance about the materials is understandable. Steel was relatively new in the days of the Tay Bridge construction and its properties had not been investigated in any detail. Iron was by then a familiar material, but who knew how an iron structure of that scale would behave? Indeed, judging by the fact that bridges and structures still collapse with distressing frequency through material failure, we apparently have a few things still to discover.

Apart from the consideration of the materials, many disasters were caused simply by an ignorance of the concepts and applications of the basic laws of mechanics. However, many of these concepts and engineering subjects and design techniques have only been acquired and integrated by virtue of the great and small disasters of the past, and the miseries of failed engineering designers who were obliged to design in ignorance. This we should remember: it will probably apply to us, too, in our time.

The Course Team dares to hope that when you have completed this very introductory course, you will appreciate the relevance and usefulness of a knowledge of general engineering mechanics. You may decide to study more in this field, or you may not. But at least we hope that by then you will be familiar with what is involved, and that you will understand better the potentials and limitations of engineering science.

1.2 Meanwhile, in the Pacific

Easter Island lies in the middle of the Pacific Ocean between South America and New Zealand. It was discovered on Easter Day in 1722. Dotted all over the island stood hundreds of huge stone statues 10 metres or more high. Each had a great stone sphere, or top-knot, balanced on its head like a crown.

Figure 8   The Easter Island statues in 1973

Captain Cook went to Easter Island in 1772 and found the statues still standing. They were set on platforms of stone and the blocks of stone were cut and polished and fitted together neatly without cement or mortar. Captain Cook thought then that the statues and their platforms must have been very old. Today most of the statues have toppled off their platforms; some are buried up to their heads in the ground while others lie on their faces or their backs.

The stone for the statues came from a quarry in the crater of an extinct volcano at one end of the island. The top-knots of special red stone came from the other end of the island. No one could think how in ancient times the statues had been carried to their places all over the island, set upright on their platforms, and crowned with their top-knots. The top-knots must have been lifted on separately. There were no cranes, and apparently there never had been anything to make such machines with. There was no metal and not much wood grew on the island; what trees there had been were stunted.

In 1956 the mayor of the village showed a certain explorer, Thor Heyerdal, how the people of the island had managed. The mayor had never lifted a statue before, but his father had told him how it was done. His father had been told by *his* father, and so on back through eleven generations from the last man who had taken part.

The mayor and eleven other men went to work on a stone man lying in front of the platform from which it had fallen. It was lying on its face. They took long strong poles and collected a great many stones of all sizes. First they pushed two poles under one side of the statue and levered it up. As fulcrums for the levers they used big stones. Four or five men pulled on each pole. When the side of the statue lifted a little, other men pushed stones in underneath it. They used small stones at first. Next they levered up the statue a little on the other side and put stones under that, too. Then they went back to the first side and put more stones under there. Soon they could get big stones beneath it.

The stones had to be fitted together so that they would take the weight of the statue. Some of them cracked like lumps of sugar, but at the end of a day's work the statue lay on a tower of stones one metre high. So it rose slowly up, until after nine days it lay four metres above the ground. By then the men could not reach the ends of the poles to pull on them so they attached ropes to these ends.

After that they went on building up stones under the giant's face and chest, but put no more under its lower end. So it tilted slowly, with its lower end at the level of the platform, its head high above it: it had to be steadied with ropes from its forehead. With a last pull on the levers the giant slid down the pile of stones and tipped on to its platform. It wobbled, but then settled.

In this way twelve men using levers and a slope of stones raised a statue weighing 30 tons through a height of 4 metres, and set it upright. They took 18 days.

The stone statues could be moved long distances across the island on sledges made from tree trunks. The mayor demonstrated this. The topknots of red stone were about as high as a man and weighed up to 10 tons. They were brought round the coast on canoes of a special sort. (We know the way they came because some dropped off the canoes and still lie in the water.) They were trundled up the beach on rollers, and lifted on to the heads of the statues using towers of stones. All the stones were then taken away. Thus hundreds of years later nobody could imagine how it had been done.

Twelve men lifted 30 tons 4 metres up in nine days. They used their own power, with levers and many stones as a machine. They did a great deal of work but it took them some time. They were not *powerful*, but they were energetic.

Engineers can see many applications of the laws of mechanics in such an example. The raising of the statue brings in considerations of equilibrium, energy, the strength and properties of materials, forces, work, and power. Indeed, the whole example shows how much can be done using a very simple *machine*.

Figure 9   Raising the statue

1.3 Machines and power

The study of machines forms a substantial part of engineering mechanics and engineering in general, and you will find that most of the subjects you study during this course can be and are applied to the analysis and design of machines.

Machines do work. They have power: the power of a machine is the rate at which it does work. Originally this was measured in horsepower, which was quite natural since in the past the horse was the most freely available 'machine'.

Figure 10   Mill horses at work, as used in the original calculation of horsepower

One of the first to compare the power of a machine with that of horses was Thomas Savery. In 1702 he claimed that one of his engines would raise as much water as two horses working together. Later that century, the engine builder James Watt began to use information supplied to him by millers about their horses to describe the power of his engines. He was told that if a mill horse walks in a circular path 24 feet in diameter, it completes two and a half turns a minute. He took the pull of the horse to be 180 pounds, and thus found that it works at a rate of about 33 000 foot-pounds force per minute. His firm then quoted their engines as being, for example, 'fifteen horse engines'.

Although he did not realize it at the time, Watt had established a unit which was to come into world-wide use. Later the SI unit for power was named after him. One horsepower is equivalent to 745.7 watts (W) in SI units. A power of 1000 W or 1 kilowatt (kW) is equivalent to 1.341 horsepower.

These days, power outputs for most small and medium range engines and machines are still quoted in horsepower. The value is measured on a brake dynamometer (Figure 11). However, the use of SI units is becoming established. (At one time the Royal Automobile Club specified power outputs of car engines based on the somewhat arbitrary assumption that a petrol internal combustion engine of 100 cubic centimetres capacity produced one horsepower).

A machine, then, does work but this has to be channelled towards doing something useful and specific. Each machine is for a special purpose: such as making sheet metal pressings, lifting loads or flying.

Figure 11   A brake dynamometer

Simple machines like levers and pulleys can be integrated in more complex devices. Machines need power to drive them, but people worked the earliest ones themselves. Big things could be done with these early machines, as on Easter Island, but this took a long time and many men. Such jobs are now done with engines and cranes, but a great many man-powered machines are still used. At the other end of the scale there are complex machines (perhaps more mechanism than machine), which integrate on the basis of logic and memory and perform intricate functions of control. Examples are tape-controlled machine tools and automatic aircraft landing systems.

Sometimes the matching required between a human operator and a machine is so sophisticated that the machine could never fully replace the operator: the demand made on the operator is simply too complicated to design into a machine-type operation.

1.4   In parenthesis

The engineering product, be it structure, mechanism, dynamical system or energy-converting machine, is primarily supposed to be for human aid. In this lies the first of the dangers of engineering practice.

Much as we may be motivated by the scope of engineering discovery, we may be only too well aware that the development of more, bigger,

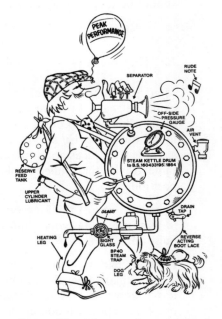

Figure 12   The matching required between machine and operator . . .

17

better technological products can be to the detriment of the quality of life. A product may not even suit the need of any one group of people. It may just be a 'technical excess' indulging those who designed, made, advertised, and sold it.

So design has a deep human meaning, and a thorough engineering training is by itself hardly an adequate springboard from which an engineer can be sure that he is going to fulfil his obligations to humanity. On the other hand, there is not much physically that can be effected in an industrial society without widely educated engineers.

There is also another point to be made. For a long-time after the original Easter Island enterprises, man harnessed beasts, other men, wind and water to do his work. With the advent of energy-converting machines, which relied particularly upon the earth's limited supply of fossil fuels, the larger-scale use of animals and manpower (in the technical sense of that word) diminished. Man became entranced by the opportunity to have his work done for him. Levers had amplified force or motion, but the new steam-age engines and electric machines were energy and power amplifiers. He never looked back. He effected bigger works and greater and greater energy transformations. He lost touch with the *scale* of energy, for energy seemed plentiful and ludicrously cheap. Energy as a resource he took for granted, depended utterly upon; he was angry if it was not delivered on time, but gave no other thought to it. It was a product of the apparently liberating science/art of engineering.

Ultimately it dawned upon some people that these energy resources were finite, even that their exhaustion might already be in view. Similarly, other consequences of large-scale energy conversion and engineering industrialization suddenly appeared to some to be daunting if not damning. It seemed that these had produced a potentially disastrous situation which might one day blight most of the planet's environment. It was no comfort to postulate, with some justification, that it was not the practice of engineering but other arts or professions of modern industrialized society that were principally to blame.

Useful and exciting as it may be, engineering by itself is neither good nor bad, merely prone to manipulation. At best, it is subservient to the needs of people. To direct it wisely, to contain its potential for long-term destruction and to make the most of its capacity for producing stable technological and environmental systems, perhaps engineering should be taught in as wide a social and ecological context as possible.

1.5 How the laws of mechanics have been built up

Men needed to know exactly how forces worked before these forces could be made to drive their big machines. Archimedes, Galileo and many others found out a great deal about how forces worked. Newton found out more. He saw how all this knowledge could fit together into a whole if his ideas about nature were correct. For example, perhaps the same gravity that pulled apples from trees could also keep the moon in position.

Newton's way of thinking out his law of gravitation and making sure it was true is a good illustration of the way in which science is built up.

First he had thought of the 'law', and saw that if it were true it could explain how the moon goes round the earth. When he worked it out, however, it seemed that the earth's pull was too strong to be the one that kept the moon circling as it does. His idea was clearly wrong. He put it aside. Sixteen years later the size of the earth was measured more closely. This had come into his reckoning, so he did the calculation again with the new size. This time he found the pull of gravity on the moon *could* account for its movement round the earth.

Next he showed that if his 'law' were true, the moon would be pulled out of this path by the sun and the other planets at certain times: he worked out exactly how much it would stray. Then he looked at the moon to see what it did do. It strayed as he had foretold. He worked out other effects that his idea of gravitation would have on the moon and planets, and again his figures proved to be right.

Thus Newton decided that his 'law of gravitation' was true, or at least had not yet been contradicted. If the moon had not moved exactly as he had reckoned, he would again have turned down his idea as wrong. He would have looked for another idea that could explain the moon's movements exactly, and tested that. He only invented the theory to explain the facts. If it did not, it was no use. In fact his idea about gravitation turned out to explain a great many more things than the movement of the moon. Among them it showed how the tides worked, and explained the shape of the earth.

In doing all this, Newton was demonstrating a *way* of finding out about natural things. By using it, the discovery of knowledge about natural things (i.e. natural philosophy) has exploded in the last 300 years.

1.6 Summary

Engineering mechanics is about motion, and how to induce it, or how to prevent it. It is concerned with the relationship between force and motion, and the sorts of motions that some forces can excite. It is also about machines and power, and about bridges and other structures: how men put together their ideas with the materials they have at hand to produce devices which serve them. The structures and machines may be separate, or they may be intimately combined to form other more sophisticated machines, such as internal combustion engines and turbines, alternators and motors. Engineering mechanics is particularly concerned with the design of all machines and structures.

The subject, then, seems to cover about all you can think of that man uses to help him do things and do work. We cannot go far in one course such as this, but it is possible to present to you the basic physical ideas and principles involved; also to show you something of how they relate to each other and have their use in the man-made world.

Even before this, we must revise and introduce some concepts and simple techniques in elementary engineering mathematics, only to help you to describe more explicitly and to understand more clearly some of the physical models on which engineering depends.

Part II

MATHEMATICS FOR MECHANICS: INTRODUCTION

Getting to feel at home with mathematical symbols is very much like learning to swim. It is not difficult and does not call for unusual intelligence. The main requirement is sufficient confidence to restrain the rising feeling of panic and to remain calm while lifting your feet off the firm bottom of verbal ability and finding that you can at least float. The difficulty of screwing up this amount of courage becomes greater the longer you spend convincing yourself that you will never be able to do it. In fact, once somebody has found that he can float you do not have to teach him to swim; all he needs is a chance to watch what other people do and to have new strokes shown to him from time to time. Style, if it is wanted (and people can enjoy themselves in a swimming pool without being oppressed by the thought that they will never make the Olympics) can be developed later. The most important thing always to bear in mind is that you must resist the temptation to expect difficulties. If you thrash about in anxiety you are bound to find yourself going under.

Thus wrote Professor Michael Hussey in his booklet 'Using Letters instead of Numbers'*, and it is a passage well worth recalling.

Mathematics is part of the language of engineering mechanics. Without it, it would be impossible to express anything quantitatively. We would only be able to describe things with words, and that is insufficient for our purposes.

You will need to be familiar with the use of algebra, trigonometry and calculus to fully appreciate the ideas and techniques of mechanics. If you do not consider yourself any sort of a mathematician, do not worry. None of the Course Team consider themselves as mathematicians either! All you require is the confidence to 'float', which this section should give you. In the subsequent units you will have the 'chance to watch what other people do' and to have 'new strokes' shown to you from time to time.

So blow up your water-wings and read on.

2.1 Symbols and their meaning

Symbols are used in mathematics simply to make it easy to write statements and descriptions in an abbreviated form: mathematical symbols have exactly the same purpose as shorthand symbols. We need only a few to be going on with and a list of these follows. You will no doubt be familiar with most or all of them but whether you are or not, a few seconds scanning over them will be a few seconds well spent.

* *The Open University* (*1971*) T100 The Man-made World: Technology Foundation Course, Unit 4S 'Using Letters instead of Numbers', *The Open University Press.*

Symbol	Meaning
$x = y$	x is equal to y
$x \neq y$	x is not equal to y
$x \simeq y$	x is approximately equal to y
$x > y$	x is greater than y
$x \gg y$	x is very much greater than y
$x \geqslant y$	x is greater than or equal to y
$x < y$	x is less than y
$x \leqslant y$	x is less than or equal to y
$x < y < z$	y is greater than x but less than z
$x \leqslant y \leqslant z$	y is greater than or equal to x but less than or equal to z
$x \rightarrow y$	x approaches the value of y
$x + y$	x is added to y
$x - y$	y is subtracted from x
$x \pm y$	y is added to or subtracted from x
$x \times y$	x multiplied by y
xy	x multiplied by y
$\dfrac{x}{y}$	x divided by y
$x \propto y$	x is directly proportional to y
$x \propto \dfrac{1}{y}$	x is inversely proportional to y
∞	infinity
π	a constant (pi), approximately equal to 3.1416, which is the ratio of the circumference of any circle to its diameter
$y = f(x)$	y is a function of x (and not equal to f times x): for instance the area of a circle A depends entirely on its diameter so you can say that A is a function of its diameter d, that is, $A = f(d)$
$y = f(x, z \ldots)$	y is a function of x and of z etc.

2.2 A reminder about the use of brackets

Brackets are used when we want to perform a simple mathematical operation, such as addition or multiplication, with a quantity or expression.

The following paragraphs are designed to remind you of the use (or abuse) of brackets in some simple operations.

1 When a positive sign precedes a pair of brackets, the brackets may be removed without changing the signs of the terms inside them. For example

$$A + (B + C - D) = A + B + C - D. \tag{1}$$

2   When a negative sign precedes a pair of brackets, the brackets may be removed if the signs of all the terms inside them are changed. For example

$$A-(B+C-D) = A-B-C+D. \tag{2}$$

3   When an expression in a pair of brackets is to be multiplied by another expression, then each term within the brackets must be multiplied by it. For example

$$A(B+C) = AB+AC \qquad \text{(and } not \ AB+C\text{)}. \tag{3}$$

4   If all the terms within a pair of brackets contain a common factor, that factor may be placed outside the brackets as a common multiplier. For example

$$(AB-AC+DA) = A(B-C+D). \tag{4}$$

5   If the expression inside a pair of brackets is to be squared or is to have some other operation performed upon it, first carry out any operations within the brackets. For example

$$(4+6)^2 = (10)^2 = 10 \times 10 = 100, \tag{5}$$

(and *not* $(4+6)^2 = 4^2+6^2 = 16+36 = 52$).

6   When an expression contains pairs of brackets within other pairs, perform the operation required to remove the innermost brackets first and then work outwards. For example

$$10\{25-[(2 \times 6)+9]\} = 10[25-(12+9)] = 10[25-21] = 10 \times 4 = 40. \tag{6}$$

There are a few pitfalls in the last example. If you are not completely sure about it, read it through carefully again and then try the following SAQ.

SAQ 1   **SAQ 1**

Write out equivalents of the following expressions, removing all the brackets:

(a) $6A(12B+3C)$,

(b) $(A+B)(A-B)$,

(c) $3A[B+C(2D+B)]$,

(d) $A-B\left[\dfrac{C}{D}+\left(E-\left(\dfrac{A}{C}\right)^2\right)C\right]$.

23

INDICES

3.1 Powers

It is useful to have a simple way of writing repeated factors. The method used is called 'raising to a power', and it works like this:

$$A \times A = A^2;$$
$$A \times A \times A = A^3;$$
$$A \times A \times A \times A = A^4;$$

and so on. Therefore in general you can see that

$$\underbrace{A \times \cdots \times A}_{n} = A^n. \tag{7}$$

The superscript n is called an *index* and its value is often called the *power* to which A is raised.

index
power

Looking at the above equalities, you can see that in moving down from one line to the next, the left-hand side is multiplied by A while the index on the right is increased by one. Moving two lines involves two multiplications by A and an increase of two in the index.

Suppose that we wished to multiply the first line by the third. We would get

$$(A \times A) \times (A \times A \times A \times A)$$
$$= (A \times A \times A \times A \times A \times A) = A^6.$$

The index of A^6 is the sum of the two indices of A^2 and A^4 (first and third lines), that is

$$A^2 \times A^4 = A^6 = A^{(2+4)},$$

and in general

$$A^n \times A^m = A^{(n+m)}. \tag{8}$$

Another rule can be obtained by looking, say, at the third line and dividing it by the first.

$$\frac{A \times A \times A \times A}{A \times A} = A \times A$$

which is clearly A^2. But the index of A^2 is equal to the difference between the indices of A^4 and A^2, that is

$$\frac{A^4}{A^2} = A^2 = A^{(4-2)},$$

and in general

$$\frac{A^n}{A^m} = A^{(n-m)}. \tag{9}$$

This rule reveals something else that is not so readily seen from the original equalities. The process of dividing by A on the left and reducing the index by one on the right could be carried backwards indefinitely. Let us do this and see what we get.

$$A \times A \times A \times A = A^4,$$
$$A \times A \times A = A^3,$$
$$A \times A = A^2,$$
$$A = A^1,$$
$$\frac{A}{A} = 1 = A^0,$$
$$\frac{A}{A \times A} = \frac{1}{A} = A^{-1},$$
$$\frac{A}{A \times A \times A} = \frac{1}{A^2} = A^{-2},$$
$$\frac{A}{A \times A \times A \times A} = \frac{1}{A^3} = A^{-3}.$$

and so on. Therefore in general

$$\frac{1}{(\underbrace{A \times \cdots \times A}_{n})} = \frac{1}{A^n} = A^{-n}. \tag{10}$$

These results give a neat way of expressing repeated multiplication and repeated division by A, and A can of course be any number or expression.

There is one thing that you may have noticed in working through this: although A stands for any unspecified number (positive or negative), A^0 is equal to unity – a quite definite value that does not depend on A. So any number raised to the power of zero is equal to unity.

3.2 Roots

A common question in the numbers game is 'what number multiplied by itself gives A?' The number that gives the answer to the question is called the *square root* of A and the most common way of writing this is $\sqrt[2]{A}$. A piece of shorthand again.

square root

We could equally well ask the question 'what number multiplied by itself twice gives A?' In this case the answer is the *cube root* of A, written $\sqrt[3]{A}$.

cube root

In general then there exists an infinite number of roots of any number and the nth root (where n is any number) is written $\sqrt[n]{A}$.

***n*th root**

Now look at what happens when we want to multiply or divide two or more different roots of the same number. For instance suppose we wished to multiply the square root of A by the fourth root of A. We would have, in our present notation,

$$\sqrt[2]{A} \times \sqrt[4]{A}.$$

How do we express the result? There is a way of doing so using the $\sqrt{}$ notation, but there is a much neater and clearer notation which uses indices.

If we square A^n, we get, using our previous rule of equation (8),

$$(A^n)^2 = A^n \times A^n = A^{n+n} = A^{2n}$$

and in general

$$(A^n)^m = A^{(n \times m)}. \tag{11}$$

So we see that squaring a power of A doubles the index, which suggests, going backwards, that taking the square-root of A involves halving the index. Let us try it and see.

We know

$(\text{square-root of } A) \times (\text{square-root of } A)$ must equal A.

Thus we let

$(\text{square-root of } A) = A^x$ and try to find x.

Substituting gives

$$A^x \times A^x = A = A^1.$$

So following the rule of equation (8), x must equal $\frac{1}{2}$ because

$$A^{1/2} \times A^{1/2} = A^{(1/2+1/2)} = A^1 = A.$$

We can thus write the square root of A as $A^{1/2}$, and in general the nth root of A is written as $A^{1/n}$. The result of multiplying $\sqrt[2]{A}$ and $\sqrt[4]{A}$ is found simply by rewriting the product as

$$A^{1/2} \times A^{1/4}$$

and this, using our rule (8) becomes

$$A^{(1/2+1/4)} = A^{3/4}.$$

In this case you can see that the resulting index is not a whole number, so what does $A^{3/4}$ mean?

If the fractional index is separated into its numerator and denominator thus:

$$A^{3/4} = A^{3 \times 1/4},$$

we can see by looking at the rule of equation (11) that this must mean $(A^3)^{1/4}$, or $(A^{1/4})^3$ which is the same. Thus the piece of shorthand $A^{3/4}$ stands for the fourth root of A cubed. Using the $\sqrt{}$ notation it would be written $\sqrt[4]{A^3}$.

There are thus *fractional indices* which involve both the root *and* the power of a number. The numerator is the power to which the number is raised while the denominator is the appropriate root of the number.

fractional indices

Finally consider the signs of the numbers involved. If $A^{1/2}$ is a positive number, then squaring it gives A, a positive number. If $A^{1/2}$ is a negative number, squaring gives A, again a positive number. We cannot, by squaring any real number, positive or negative, produce a negative number. So negative numbers cannot have real square roots; however, they can have real negative cube, fifth or seventh, that is real odd-numbered roots because when the latter are all multiplied together the result is a negative number. (A negative number can have even-numbered roots which are 'unreal' or 'imaginary'. You will come across this in Section 6.3.)

An important point to be noticed is that all positive numbers have *two* square roots, one positive and one negative since in each case multiplication by itself produces a positive number.

Thus if

$$A = x,$$

then

$$A^{1/2} = \pm\sqrt{x}. \tag{12}$$

SAQ 2

SAQ 2

(a)   Express with positive indices

(i) $\dfrac{2x^{-3}y^2}{7z^{-4}y}$,   (ii) $\dfrac{\sqrt[3]{y^{-4}}}{\sqrt{y^2}}$.

(b) Evaluate   (i) $\left(\dfrac{81}{16}\right)^{3/4}$,   (ii) $(64)^{-3/2}$.

(c) Simplify $\dfrac{(x^4yz^{-3})^2 \times (x^{-5}y^2z)}{(xz)^{1/2}}$.

3.3 Logarithms

Having looked at powers and roots you should now be familiar with terms such as A^n (refer to equation (11) if you need to). The number n is called the *power* to which the *base* A is raised. The value Q of a number base raised to some numerical power is given by

$$A^n = Q. \tag{13}$$

A logarithm is simply a device that is used to convert numbers into a base and its powers. In fact, the logarithm of the positive quantity Q to base A is defined as the power to which A must be raised to make it equal to the given quantity Q. From equation (13) this is clearly the power n. Hence the logarithm of Q to the base A is equal to n. This statement is written

$$\log_A Q = n. \tag{14}$$

The two statements (13) and (14) are equivalent. They express the relationship between A, n and Q, and by substituting one in the other we have

$$A^{\log_A Q} = Q. \tag{15}$$

The logarithms used in everyday calculations are those with a base of 10 and these are the *common logarithms* that you will find on pages 2 and 3 of the book of logarithms for this course.* In these tables, five figures are placed opposite each of the numbers from 10 to 99. These five figures are called the *mantissa* and are the decimal part of the logarithm. The *characteristic*, which may be either positive or negative, is the integral part and has to be supplied when writing down the complete logarithm of any given number. This can be determined for logarithms to the base 10 in the following way.

power, base

common logarithms

mantissa
characteristic

Since
$$A^n = Q,$$

for the base

$$A = 10.$$
$$10^n = Q.$$

We can therefore write

$$\log_{10} Q = n.$$

Also:

$$1 = 10^0, \quad \log_{10} 1 \;\; = 0;$$
$$10 = 10^1, \quad \log_{10} 10 \;\; = 1;$$
$$100 = 10^2, \quad \log_{10} 100 = 2.$$

In the section on indices you saw that 0.1 or $\frac{1}{10}$ can be written as 10^{-1}, also 0.01 or $\frac{1}{100}$ can be written as 10^{-2}. Hence

$$\log_{10} 0.1 \;\; = \log_{10}\left(\tfrac{1}{10}\right) = -1,$$
$$\log_{10} 0.01 = -2, \text{ etc.}$$

Instead of writing the negative sign in front of the characteristic, it is

---

* Castle, F. (1972) Five Figure Logarithmic and other Tables *Macmillan Education Ltd.*

customary in logarithms to place it over the top; thus, $\log_{10} 0.1$ is not usually written as -1 but as $\bar{1}$, and $\log_{10} 0.01 = \bar{2}$.

The mantissa consists of a series of numbers. Thus

$$\log_{10} 1 \quad = 0.000\,00,$$
$$\log_{10} 10 \quad = 1.000\,00,$$
$$\log_{10} 100 = 2.000\,00, \quad \text{and so on.}$$

These mantissas are all equal to zero but as $\log_{10} 1$ is $0.000\,00$ and $\log_{10} 10$ is $1.000\,00$, it is evident that the logarithm of all numbers between 1 and 10 will consist of a certain number of decimals following a characteristic of zero.

$$\text{Log}_{10} 2 = 0.301\,03$$

indicates that the characteristic of the logarithm is 0 and the mantissa is .301 03. It also shows that if we raise 10 to the power 0.301 03 we shall obtain 2, or that

$$10^{0.301\,03} = 2.$$

Looking in the logarithmic tables opposite the number 47, for example, we find the mantissa to be .672 10 and as 47 lies between 10 and 100, the characteristic is 1. Hence

$$\log_{10} 47 = 1.672\,10.$$

The number 470 lies between 100 and 1000 and therefore the characteristic in this case is 2. Hence

$$\log_{10} 470 = 2.672\,10.$$

Similarly, the logarithms of 4700 and 47000 are 3.672 10 and 4.672 10 respectively; in each case the mantissa is the same but the characteristic is a number one less than the number of figures to the left of the decimal point.

Exercise

Write down $\log_{10} 0.047$.

Here one zero follows the decimal point: $\log_{10} 0.047$ lies between $\log_{10} 0.1 = -1$, and $\log_{10} 0.01 = -2$. Thus the characteristic is $\bar{2}$. From the tables the mantissa of 47 is .672 10. Therefore

$$\log_{10} 0.047 = \bar{2}.672\,10.$$

It is important to remember that *the mantissa is always positive*: the logarithm in the exercise is really made up of two parts, namely -2 and $+.672\,10$. Strictly speaking

$$\log_{10} 0.047 = -2 + .672\,10 = -1.327\,90.$$

SAQ 3

From your tables find the common logarithm of the following:
(a) 0.0063, (b) 548, (c) 2.17, (d) 43.6, (e) 0.018, (f) 0.000 27.

Instructions on how to find logarithms of numbers are also given on page 61 of Castle's book of tables. You may find these helpful.

Exercise

Given the logarithm of a number is 2.472, find the number.

Looking at the tables of antilogarithms on pages 4 and 5, opposite the mantissa of .47 we have 29 512. Moving to the right until you are in the column headed 2 the mantissa becomes .296 48. The characteristic is 2 so the number must be 296.48.

SAQ 4

Find the number whose logarithm is:
(a) 2.189 67, (b) 0.635 33, (c) $\bar{1}$.293 76, (d) $\bar{4}$.941 89, (e) 1.495 91, (f) $\bar{1}$.111 79.

In Castle's book of tables, on pages 36 and 37, you will find some other tables called hyperbolic or *Naperian logarithms* (after the discoverer of logarithms, Napier). These are logarithms to the base 2.718 28 (to five decimal places). This number has special significance in mathematics and in the application of mathematics to engineering and science. It is denoted by the letter e. Basically e is the number which is the sum of the series

$$\frac{2}{1} + \frac{1}{1 \times 2} + \frac{1}{1 \times 2 \times 3} + \frac{1}{1 \times 2 \times 3 \times 4} + \cdots$$

This is discussed more fully in Appendix A.

Logarithms can be used to perform operations such as multiplication, division, raising to any power, finding the root of a number, and so on. This follows simply from use of the laws of indices.

Suppose we have two numbers M and N. From equation (15) and using a base of 10,

$$M = 10^{\log_{10} M}$$

and

$$N = 10^{\log_{10} N}.$$

Thus

$$M \times N = 10^{\log_{10} M} \times 10^{\log_{10} N},$$

and from equation (8)

$$M \times N = 10^{\log_{10} M} \times 10^{\log_{10} N} = 10^{(\log_{10} M + \log_{10} N)}.$$

Hence from equations (13) and (14)

$$\log_{10}(M \times N) = \log_{10} M + \log_{10} N. \tag{16}$$

Thus if the logarithm of the *product* of two numbers is required, it is only necessary to *add* the separate logarithms.

Exercise

Using logarithms, multiply 0.2885 by 0.9150.

From the tables,

$$\log_{10} 0.2885 = \bar{1}.460\ 15$$

and

$$\log_{10} 0.9150 = \bar{1}.961\ 42.$$

These two logarithms have to be added, but remember that the mantissa is always positive. This may be clearer if we write

$$\log_{10} 0.2885 = -1 + .460\ 15$$
$$\log_{10} 0.9150 = -1 + .961\ 42$$

and adding gives $\quad -2 + 1.421\ 57 = \bar{1}.421\ 57.$

The answer to 0.2885×0.9150 is now given by the antilogarithm of $\bar{1}.421\ 57$. From the tables this is 2640, and since the characteristic is $\bar{1}$ the answer must be 0.2640.

SAQ 5

Using logarithms, evaluate the following products:
(a) $30.98 \times 0.002\,58$;
(b) $0.000\,215 \times 349\,000\,0$;
(c) $3.54 \times 0.026 \times 1.34$.

For *division* equation (9) shows that the logarithms must be *subtracted*, so we can write

$$\log_{10}\left(\frac{M}{N}\right) = \log_{10} M - \log_{10} N. \qquad (17)$$

SAQ 6

SAQ 6

Using logarithms, evaluate the following quotients:

(a) $\dfrac{0.006\,362}{2.052}$;   (b) $\dfrac{99.94}{2890}$;   (c) $\dfrac{0.5}{0.0065}$.

The logarithm of a positive quantity raised to a power can be found using equation (11). This results in the identity

$$\log_{10}(M^p) = p\log_{10} M, \qquad (18)$$

and if

$$p = \frac{1}{r}$$

then

$$\log_{10}(M^{1/r}) = \frac{1}{r}\log_{10} M. \qquad (19)$$

thus the logarithm of the pth power of a positive quantity is p times the logarithm of the quantity, and the logarithm of the rth root of a positive quantity is $1/r$ times the logarithm of the quantity.

Exercise

Evaluate the following:

(a) $(0.07)^3$,   (b) $(475)^{1/3}$.

(a) From the tables

$$\log_{10} 0.07 = \bar{2}.845\,10$$
$$= -2 + .845\,10.$$

Using equation (18)

$$\log_{10}(0.07)^3 = 3 \times \log_{10} 0.07 = 3 \times (-2 + .845\,10)$$
$$= -6 + 2.535\,30$$
$$= \bar{4}.535\,30.$$

The antilogarithm of $\bar{4}.535\,30$ is $0.000\,343$. Hence

$$(0.07)^3 = 0.000\,343.$$

(b) From the tables

$$\log_{10} 475 = 2.676\,69.$$

Using equation (19)

$$\log_{10}(475)^{1/3} = \tfrac{1}{3} \times \log_{10} 475 = \frac{2.676\,69}{3}$$
$$= 0.8922.$$

The antilogarithm of 0.8922 is 7.802. Hence

$$(475)^{1/3} = 7.802.$$

SAQ 7

Using logarithms, evaluate:

(a) 0.621×0.026;

(b) $0.010\,19 \times 23.04$;

(c) $\dfrac{0.000\,948\,1}{0.0157}$;

(d) $\dfrac{0.2098}{36.15}$;

(e) $\dfrac{619.3 \times 0.117}{1.43}$;

(f) $\dfrac{10 \times 0.013\,42}{0.0055}$

(g) $\dfrac{(3.1416)^{12}}{(2.1782)^{20}}$;

(h) $\dfrac{42 \times (0.0016)^{7/4}}{\sqrt[3]{108}}$.

SAQ 7

TRIGONOMETRY

4.1 Sines, cosines and tangents:

The triangle ABC (Figure 13) has one of its angles, angle ACB, equal to 90° (a 'right angle') and is therefore called a right-angled triangle. Angle BAC is equal to $\theta°$ so the remaining angle will be equal to $(180° - 90° - \theta°) = (90° - \theta°)$, since the sum of the angles of any triangle equals 180°.

Let the length AC $= a$ (since it is the *a*djacent side to the angle θ); let BC $= o$ (*o*pposite to θ), and AB $= h$ (since this side opposite the angle of 90° is called the *h*ypotenuse).

The *sine*, *cosine* and *tangent* of the angle θ are then defined as:

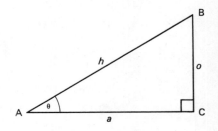

Figure 13

sine, cosine, tangent

$$\sin \theta = \frac{BC}{AB} = \frac{o}{h}; \tag{20}$$

$$\cos \theta = \frac{AC}{AB} = \frac{a}{h}; \tag{21}$$

$$\tan \theta = \frac{BC}{AC} = \frac{o}{a}. \tag{22}$$

Also

$$\frac{\sin \theta}{\cos \theta} = \frac{o}{h} \times \frac{h}{a} = \frac{o}{a} = \tan \theta. \tag{23}$$

'Sin,' 'cos' and 'tan' are the conventional abbreviations. The expressions for the reciprocal quantities *cosecant*, *secant* and *cotangent* are:

cosecant, secant, cotangent

$$\operatorname{cosec} \theta = \frac{1}{\sin \theta} = \frac{h}{o}; \tag{24}$$

$$\sec \theta = \frac{1}{\cos \theta} = \frac{h}{a}; \tag{25}$$

$$\cot \theta = \frac{1}{\tan \theta} = \frac{a}{o}. \tag{26}$$

The relationships (20) to (26) hold for all values of θ between 0° and 90°, but what are the expressions for the sine, cosine, and tangent of angles above 90°? Most doors can swing through such angles (called *obtuse* angles), so clearly these exist.

The question can be most easily answered if a set of Cartesian (X, Y) axes are superimposed on the triangle ABC (Figure 14). The point B will be at a position where its X ordinate, x (AC), is equal to a and its Y ordinate y (BC) is equal to o. θ is measured anti-clockwise from the X axis as shown.

For all values of θ within the top right-hand quadrant formed by the axes (called by convention the *first* quadrant), all values of x and y are positive, and

$$\sin \theta = \frac{y}{h} = \frac{o}{h},$$

$$\cos \theta = \frac{x}{h} = \frac{a}{h},$$

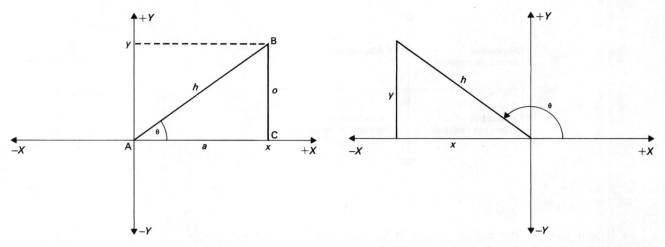

Figure 14   *Figure 15*

$$\tan \theta = \frac{y}{x} = \frac{o}{a}.$$

When θ approaches 90°, x approaches zero and y and h become equal to each other. Therefore

$$\sin 90° \rightarrow 1, \tag{27}$$
$$\cos 90° \rightarrow 0, \tag{28}$$
$$\tan 90° \rightarrow \infty. \tag{29}$$

When θ increases further into the second quadrant, y and h are still positive but x has become negative (Figure 15).

The expressions for θ then become

$$\sin \theta = \frac{y}{h} \text{ (positive)},$$

$$\cos \theta = \frac{x}{h} \text{ (negative because } x \text{ is negative)},$$

$$\tan \theta = \frac{y}{x} \text{ (negative because } x \text{ is negative)}.$$

Expressions for values of θ in the third and fourth quadrant are found similarly, by noting the sign changes of x and y as θ is increased. Try this and you will find that for the third quadrant, $180° < \theta < 270°$,

$$\sin \theta = \frac{y}{h} \text{ (negative because } y \text{ is negative)},$$

$$\cos \theta = \frac{x}{h} \text{ (negative because } y \text{ is negative)},$$

$$\tan \theta = \frac{y}{x} \text{ (positive because both } y \text{ and } x \text{ are negative)}.$$

For the fourth quadrant, $270° < \theta < 360°$,

$$\sin \theta = \frac{y}{x} \text{ (negative because } y \text{ is negative)}$$

$$\cos \theta = \frac{x}{h} \text{ (positive)},$$

$$\tan \theta = \frac{y}{x} \text{ (negative because } y \text{ is negative)}.$$

Figure 16 summarizes these changes in sign of the trigonometric functions of θ, as θ varies from 0° to 360°.

33

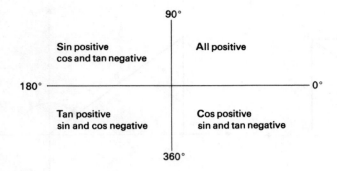

Figure 16

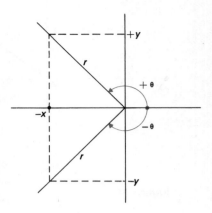

Figure 17

If you look at the capital letters in the diagram, you will see that starting from the fourth quadrant and going anti-clockwise they spell out the word **CAST**. You may find this a convenient way of remembering the relative signs.

4.2 Trigonometric ratios of some related angles

Figure 17 shows the two angles $+\theta$ and $-\theta$. Angles measured anti-clockwise (as in the previous section) are usually taken as being positive. If angles are then measured clockwise, they must be negative.

Now $\sin\theta = y/r$, so that if y is negative, $\sin\theta$ is negative, $\tan\theta$ is negative, and $\cos\theta$ is not affected.

So from the figure,

$$\sin(-\theta) = -\sin\theta.$$

Similarly

$$\cos(-\theta) = \cos\theta,$$

and

$$\tan(-\theta) = -\tan\theta. \tag{30}$$

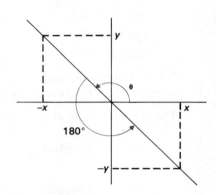

Figure 18

Now suppose $180°$ are added to θ (Figure 18). This gives

$$\sin(180°+\theta) = \frac{y}{r} = -\sin\theta \text{ (as } y \text{ is negative)}$$

Similarly

$$\cos(180°+\theta) = -\cos\theta, \tag{31}$$

and

$$\tan(180°+\theta) = \tan\theta.$$

Expressions for $(180°-\theta)$, (Figure 19), can be found from changing the sign of θ and using the results of equation (30):

$$\sin(180°-\theta) = -\sin(-\theta) = \sin\theta,$$
$$\cos(180°-\theta) = -\cos(-\theta) = -\cos\theta, \tag{32}$$
$$\tan(180°-\theta) = \tan(-\theta) = -\tan\theta.$$

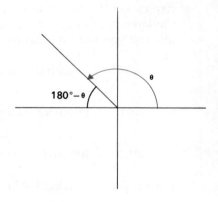

Figure 19

SAQ 8

SAQ 8

Find the equivalent trigonometric ratios in terms of θ for the following:

(a) $\sin(90°+\theta), \cos(90°+\theta), \tan(90°+\theta)$;
(b) $\sin(90°-\theta), \cos(90°-\theta), \tan(90°-\theta)$;
(c) $\sin(270°+\theta), \cos(270°+\theta), \tan(270°+\theta)$;
(d) $\sin(270°-\theta), \cos(270°-\theta), \tan(270°-\theta)$.

4.3 Graphs of sine, cosine and tangent of an angle

Plotting values of $\sin\theta$, $\cos\theta$ and $\tan\theta$ against the corresponding values of θ in degrees produces the graphs in Figures 20 (a), (b) and (c).

When θ increases beyond 360° each graph simply repeats itself, so the functions are said to be periodic. The period of both $\sin\theta$ and $\cos\theta$ is 360°, while that of $\tan\theta$ can in fact be seen to be 180°.

The graph of $\tan\theta$ may look peculiar to you because it appears to be discontinuous. Actually, at the values of 90°, 270°, 450°, etc. $\tan\theta$ is infinite. As θ approaches 90°, $\tan\theta$ increases indefinitely. When θ reaches 90° the value of $\tan\theta$ switches: for an infinitesimal increase of θ above 90°, the value of $\tan\theta$ can be said to be negative and very large. As θ increases further, $\tan\theta$ approaches zero from the negative side and gets there when $\theta = 180°$. The process is then repeated for values of θ in the next period.

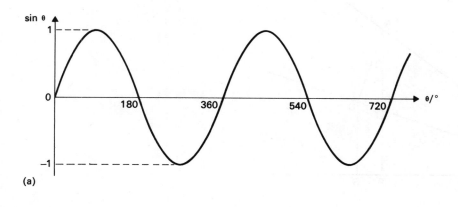

(a)

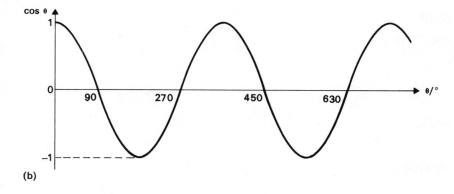

(b)

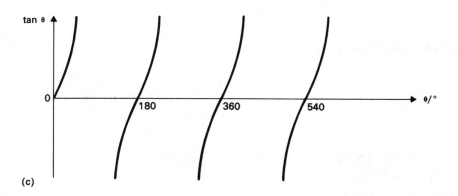

(c)

Figure 20   Graphs of (a) sin θ; (b) cos θ; (c) tan θ

35

4.4 Some useful trigonometrical indentities

An identity is an equation which is true for all values of the variable(s) for which both sides of the equation are defined. This section does not give the derivation of a large number of trigonometrical identities. However, you will need to *use* some of them from time to time during the course, so we will just derive one or two and then list the others.

To obtain an expression for the sine of the sum of two angles where the sum is an acute ($< 90°$) angle, consider Figure 21.

The angle QOP is A and the angle NOQ is B. We want to find $\sin(A + B)$. N is any point on the steepest line and angles NPO, QSO, NRQ, and NQO are right-angles ($90°$). The angle RNQ is therefore equal to A. If you do not believe this, check the angles of triangles NTQ and OTP.

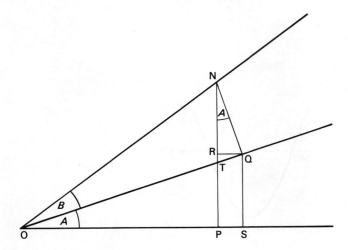

Figure 21

From the right-angled triangle NOP,

$$\sin(A+B) = \frac{PN}{ON} = \frac{PR+RN}{ON}$$

$$= \frac{SQ+RN}{ON}.$$

From the right-angled triangle OQS,

$$SQ = OQ \sin A;$$

and from the right-angled triangle OQN,

$$OQ = ON \cos B.$$

Hence

$$SQ = ON \sin A \cos B.$$

From the right-angled triangles NRQ and OQN we have

$$RN = NQ \cos A,$$
$$NQ = ON \sin B.$$

Therefore

$$RN = ON \cos A \sin B.$$

Substituting these expressions for SQ and RN back into the original expression for $\sin(A+B)$ produces the identity

$$\sin(A+B) = \frac{ON \sin A \cos B + ON \cos A \sin B}{ON}$$

$$= \sin A \cos B + \cos A \sin B. \tag{33}$$

The following identitites can be derived in a similar way:

$$\sin(A-B) = \sin A \cos B - \cos A \sin B; \qquad (34)$$
$$\cos(A+B) = \cos A \cos B - \sin A \sin B; \qquad (35)$$
$$\cos(A-B) = \cos A \cos B + \sin A \sin B. \qquad (36)$$

SAQ 9

Find the identities for $\tan(A+B)$ and $\tan(A-B)$.

SAQ 9

The identities we have considered up to now are called the *addition theorems*, and they enable a further set of identities to be found. By writing $B = A$ in these theorems, we can obtain expressions for the sine, cosine and tangent of $2A$ in terms of the sine, cosine and tangent of A.

addition theorems

From expression (33) we get

$$\sin(A+A) = \sin A \cos A + \cos A \sin A,$$

or

$$\sin 2A = 2 \sin A \cos A. \qquad (37)$$

For the cosine

$$\cos 2A = \cos^2 A - \sin^2 A. \qquad (38)$$

SAQ 10

(a) Find the identity for $\tan 2A$ in terms of $\tan A$.
(b) Use the theorem of Pythagoras (Appendix C) to show that

$$\cos^2 A + \sin^2 A = 1.$$

SAQ 10

This method can be extended for finding identities for angles equal to $3A$, $4A$, or even $\frac{1}{2}A$ and so on. You will be introduced to these as and when necessary during the course.

SAQ 11

(a) If θ is an acute angle whose sine is equal to 0.8660, find the value of the angle in degrees and the value of its cosine and tangent.
(b) If $\cos \theta = -0.5000$, find the two possible values of θ in the range $0° < \theta < 360°$.
(c) Prove that

$$\sin 330° \cos 390° - \cos 570° \sin 510° = 0.$$

SAQ 11

TRIANGLES

5.1 Notation

A triangle has six parts or elements: three sides and three angles. In Figure 22 the angles A, B and C have the sides opposite them denoted by the corresponding letters a, b and c.

The sides of a triangle are independent of one another except for the fact that the sum of the lengths of any two of them must be greater than the length of the third. The angles, however, are not independent since the sum of the angles of any triangle is 180°. Hence if two angles are known the third can be found by simple subtraction.

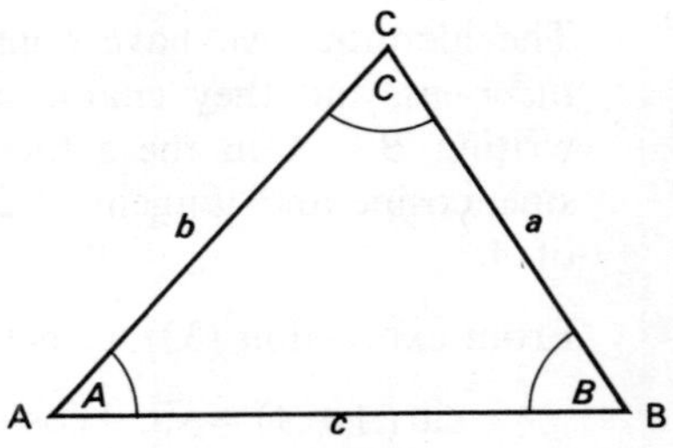

Figure 22

5.2 The sine rule

Let O be the centre of the circle circumscribing the triangle ABC (Figure 23). The line BD is a diameter of the circle and the angle BCD, being the angle in a semicircle, is a right angle (see Appendix B). Since triangles ABC and BDC have a common base BC and points A and D both lie on the circle, angle BAC will be equal to angle BDC (this is from another property of a circle).

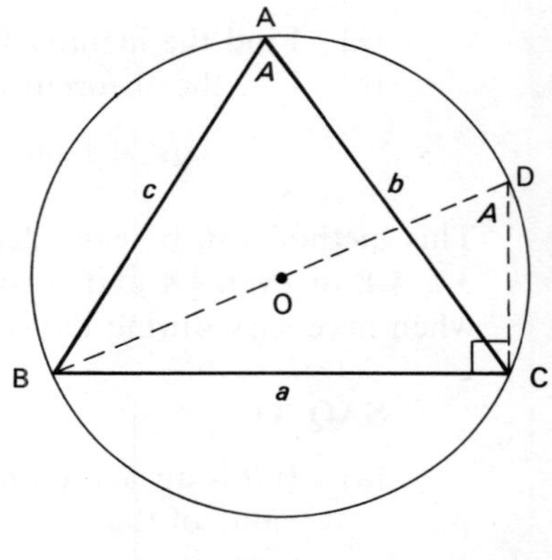

Figure 23

Hence from the triangle BDC,

$$\sin A = \frac{\text{BC}}{\text{BD}}$$

$$= \frac{a}{\text{BD}},$$

or

$$\text{BD} = \frac{a}{\sin A}.$$

By joining AD instead of DC we could prove similarly that BD = $c/\sin C$, and by starting the construction from C instead of B, we could show that BD = $b/\sin B$. These three results give the formula

$$\frac{a}{\sin A} = \frac{b}{\sin B} = \frac{c}{\sin C}. \qquad (39)$$

This is generally known as the *sine rule*. It is true for any triangle whether it includes an obtuse angle or not.

sine rule

5.3 The cosine rule

In the triangle ABC of Figure 24, BD is the perpendicular from B to the base CA. The right-angled triangle DAB gives

$$\text{BD} = c \sin A,$$

$$\text{AD} = c \cos A.$$

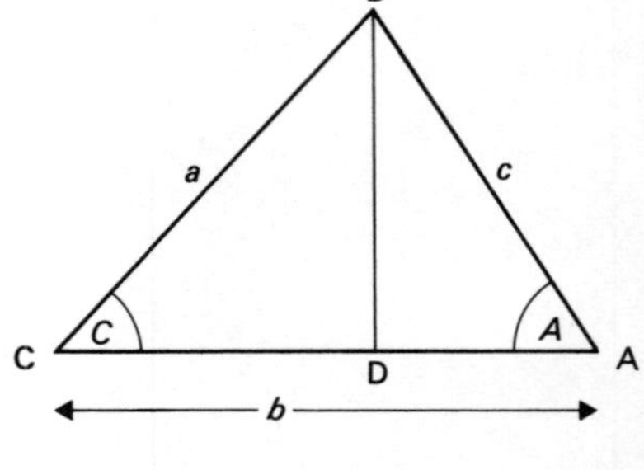

Figure 24

So

$$\text{CD} = \text{CA} - \text{AD} = b - c \cos A.$$

Applying the theorem of Pythagoras (see Appendix C) to the right-angled triangle DCB gives

$$(CB)^2 = (CD)^2 + (BD)^2$$

or

$$a^2 = (b - c\cos A)^2 + (c\sin A)^2$$
$$= b^2 - 2bc\cos A + c^2\cos^2 A + c^2\sin^2 A$$
$$= b^2 - 2bc\cos A + c^2(\cos^2 A + \sin^2 A).$$

Since $\cos^2 A + \sin^2 A = 1$,

$$a^2 = b^2 - 2bc\cos A + c^2$$

or

$$a^2 = b^2 + c^2 - 2bc\cos A.$$

The two similar formulae

$$b^2 = c^2 + a^2 - 2ca\cos B$$

and

$$c^2 = a^2 + b^2 - 2ab\cos C$$

(40)

can be derived by the same method. These expressions (40) are the *cosine rules*. They are useful in the solution of triangles when at least two sides are given.

cosine rule

5.4 The area of a triangle

The area of a right-angled triangle is given by

$$\text{Area} = \tfrac{1}{2}(\text{height} \times \text{base}).$$

Consider the triangle in Figure 25. If BD is the perpendicular from B to the base AC, then since $BD = c\sin A$, the area of the triangle is given by

$$\text{Area} = \tfrac{1}{2}(BD \times CD) + \tfrac{1}{2}(BD \times AD)$$
$$= \tfrac{1}{2}(BD \times CA)$$
$$= \tfrac{1}{2}bc\sin A.$$

(41)

In the same way it can be shown that the area also is given by

$$\text{Area} = \tfrac{1}{2}ca\sin B$$

or

$$\text{Area} = \tfrac{1}{2}ab\sin C.$$

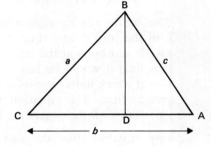

Figure 25

EQUATIONS

6.1 Linear equations in two unknowns

Suppose the number of bolts stacked in a metal tray at the start of a shift on a production line is equal to C. Throughout the shift the bolts are used at a constant rate of m per hour until at the end of the shift, eight hours later, there are p bolts left.

We can write down this information using mathematical shorthand, deriving an *equation* to describe what is happening.

We first consider the situation at the start of the shift. We can assume our time-scale begins here; that is, at this point, time $t = 0$. If we let n be the number of bolts in the tray at any time t, we can write

$$n = C \quad \text{at} \quad t = 0.$$

As our time-scale increases from zero towards 8 hours, the number of bolts in the tray decreases from $n = C$ at a rate of m bolts per hour. At any time T hours after the start of the shift, i.e. at $t = T$, the number of bolts in the tray will be given by

$$n = C(\text{ bolts at } t = 0) - m(\text{bolts per hour}) \times T(\text{hours})$$

or

$$n = (C - mT) \text{ bolts,} \tag{42}$$

and at the end of the shift

$$p = (C - 8m) \text{ bolts.}$$

Thus we have an equation which will provide us with the number of bolts n in the tray at any time T after the start of the shift. If there were insufficient bolts at the start to last through the shift, we could find out how long it would be before they ran out. We could even find out the rate at which bolts were used by counting the number of bolts left after a certain time. The equation provides us with a value of *one* of the variables if *all* the others are known.

Any equation like the one we have just derived is dimensionally *homogeneous*: since one side of an equation is equal to the other, the *dimensions* of the two sides must be the same whether these dimensions are mass, length, force, or numbers of bolts. This is simply saying that a mass cannot equal a length, and so on, but it is an important concept. It follows that each term in the equation must have the same dimensions.

homogeneous equation, dimensions

In engineering calculations the basic dimensions used are those of mass, length and time. The basic dimensions could be quantified in any system of units (SI, imperial, metric, or any other), but we will use the SI system. This system of units is compatible, which means that because the dimensions of each term in an equation are the same, so are the units. This again may seem obvious: after all, you do not add a kilogram to a mile, or a second to an hour.

Other dimensions such as force and acceleration can be written in terms of the three basic dimensions: mass, length, and time. The mass dimension is usually denoted by M, and the length dimension by L, while a period of time has a dimension of T. Sometimes M, L and T are placed in square brackets: $[M]$, $[L]$ and $[T]$. The choice of the basic dimensions is to some extent governed by convenience. It is possible to work in other dimensions such as force, length and time: [F], [L] and [T]. However this can be inconvenient, as you will see in Unit 8. Later, in the thermodynamics part of the course, we will need another basic dimension: that of temperature. This is introduced in Unit 10.

It is often possible to check that physical equations have the correct form, by inspecting the dimensions of the quantities involved. As we have seen each term must have the same units.

To check our example of the bolts, let us write in the units:

$$n(\text{no of bolts}) = C(\text{no of bolts}) - m\left(\frac{\text{no of bolts}}{\text{hour}}\right) \times T(\text{hours}).$$

The units of m have been deliberately written as a quotient. You can then see that the 'hour' in the denominator of the units of m, when multiplied by the units of T(hours), produces the unit (hour/hour). This can be thought of as a ratio whose value is 1, and a ratio can have no units. We are thus left with an equation relating numbers of bolts, each term in the equation having this unit:

$$n(\text{number of bolts}) = C(\text{number of bolts}) - mT(\text{number of bolts}).$$

The basic equation could be rearranged. For instance we could separate m from the other variables to give

$$m = \frac{C - n}{T} \text{ bolts per hour.}$$

Check the units of each term yourself to ensure that the equation is still homogeneous.

Suppose we wished to see how n varied with time T without having to calculate the value of n at every increment of time. We would want to construct a *graph* of n as it varies with time.

C and m are both constants so there are only two variables, n and T. We thus need two axes. Let us use a vertical axis for n and a horizontal axis for time T (Figure 26). We take a number of values of T and calculate the corresponding number of bolts n at those values. We then have a series of points relating to both axes through which a line can be drawn, giving us what we call the graph of n against T.

It will be easier to deal in numbers here so let us assume that at the start of the shift there are 100 bolts in the tray and they are used at the rate of 10 per hour, that is $C = 100$ and $m = 10$. Taking values of T equal to 1, 2, 3, 4 and 5 hours and using equation (42) gives the corresponding values:

$$n = 100 - (10 \times 1) = 90 \quad \text{for } T = 1;$$
$$n = 100 - (10 \times 2) = 80 \quad \text{for } T = 2;$$
$$n = 100 - (10 \times 3) = 70 \quad \text{for } T = 3;$$
$$n = 100 - (10 \times 4) = 60 \quad \text{for } T = 4;$$
$$n = 100 - (10 \times 5) = 50 \quad \text{for } T = 5;$$

showing that n decreases by 10 bolts every hour. Plotting these points on our axes and drawing a line through them gives us Figure 27.

The graph shows visually how the number of bolts in the tray varies with time. You can immediately see that n decreases at a constant rate with time. It is said to show a 'linear decrease' (since it is signified by a straight line) and the constant rate of change is called the *gradient* or *slope* of the line. In this particular case the slope is *negative* to show a *decrease* of n with time.

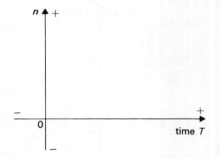

Figure 26   *Axes for the graph*

graph of a linear equation

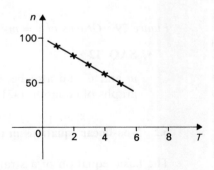

Figure 27   *Plotting the points*

gradient, slope

41

To find the complete picture we only have to extend the line to cover the time of the whole shift, as shown in Figure 28.

You can now see that the value of n where the line cuts the n axis is 100 which is equal to the initial value C. This value is called the *intercept*. The value of n at the end of the shift p can be seen to be equal to 20 bolts.

Therefore, to summarize: the particular equation

$$n = C - mT$$

can be represented by the graph of Figure 28 for values of T between 0 and 8. Here $-m$ is the slope or gradient of the line and C is the intercept on the n axis, that is it is the value of n when $T = 0$.

In the general case if a variable x varies linearly with another variable y then the relationship between them can be represented by the equation

$$y = mx + C \tag{43}$$

where m and C may be positive or negative. (In the general case we usually write them both as positive.) The graph of y against x can therefore take any one of the four forms given in Figure 29, depending on the signs of m and C.

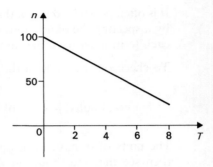

Figure 28 *Extending the line to give the intercept*

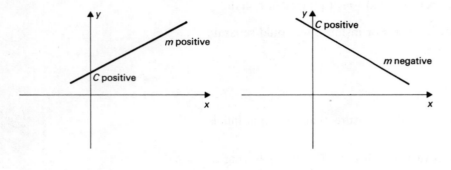

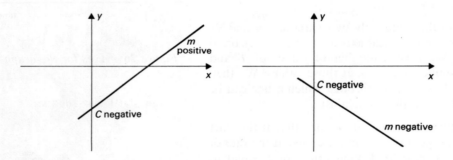

Figure 29 *Graphs of $y = mx + C$*

SAQ 12

m and C can also be zero or infinite. Sketch the corresponding graphs of equation (43) where m and C take these values.

6.2 Non-linear equations in two unknowns

The basic equation of a straight line ($y = mx + C$) is the simplest type of equation. The next most complicated is one where x is raised to some power other than 1. For example we could have

$$y = ax^2, \tag{i}$$
$$y = ax^2 + bx + c, \tag{ii}$$

or

$$y = ax^n + bx^{(n-1)} + \cdots . \tag{iii}$$

Can you see that equation (i) is the same as equation (ii) except that b and c are both zero? The general equation (iii) covers all possibilities concerning powers of x and values of the constants.

Probably the most important non-linear equation you are likely to meet for the time being is equation (ii). If you assumed values of a, b and c and plotted y against x you would get a curve which has the general shape of that in Figure 30. It is called a *parabola*.

parabola

The location of the curve relative to the x and y axes depends on the values of a, b and c. As an example, suppose $a = +1$, $b = 0$ and $c = 0$. The parabola would be located as shown in Figure 31. The equation for this situation is $y = x^2$. You can see that $y = 0$ at $x = 0$, and that y increases rapidly as x increases positively or negatively.

SAQ 13

Sketch the parabola in its correct location if the values of a, b, and c are:

(a) $a = -1$, $b = 0$, and $c = 0$;
(b) $a = +1$, $b = 0$, and $c = +1$;
(c) $a = +1$, $b = +1$, and $c = 0$.

What does a parabola represent?

In the last section we saw that a straight line on a graph represented a constant rate of change of a quantity. In the example we used it was the rate of use of bolts. Just from looking at the parabola, it appears that the slope of the curve is continually changing, that is the rate of change of a quantity is continually changing. Perhaps it changes at a constant rate!

Let us check it by measurement and see if it does. We will take a simple parabola where $a = +\frac{1}{2}$, $b = 0$, $c = 0$ as in Figure 32, and measure the slope at some arbitrary points. You may say that a curve has not got a slope; only straight lines have slopes. It is true that a parabola (or any curve) does not possess a *constant* slope but we *can* express the instantaneous value of the slope at any point on it.

SAQ 13

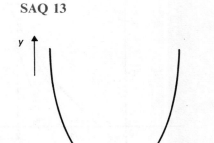

Figure 30   A parabola

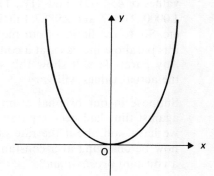

Figure 31   The parabola $y = x^2$

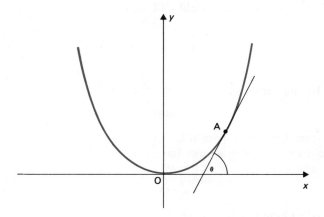

Figure 32   Slope of the curve at point A

Consider Figure 32. We can express the slope of the curve at any point A by drawing a line through A tangential to the curve and measuring the angle this line makes with the y or x axis. Figure 33 illustrates what is meant by a *tangential line*. The slope of this line can be expressed by the tangent of the angle the line makes with the x axis. This angle is θ in Figure 32.

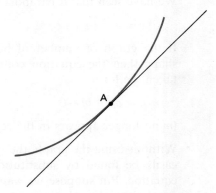

Figure 33   A tangent to a curve

tangential line

43

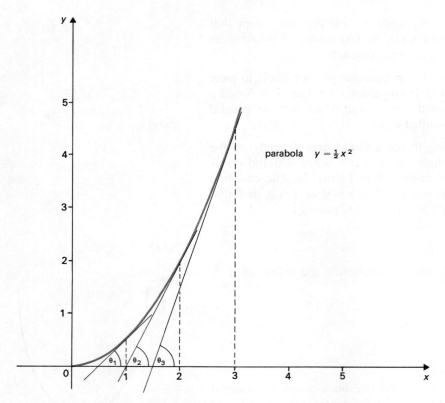

Figure 34   Slopes of the parabola $y = \frac{1}{2}x^2$

Figure 34 shows that for values of x of 1, 2 and 3, the angle θ takes values of $45°$, $63\frac{1}{2}°$ and $71\frac{1}{2}°$. The tangents of the angles are respectively 1.0000, 2.0057, and 2.9887. Further values could be measured for $x = 4$, 5, etc. So, to the limits of our measurements, we can say that the slope of this parabola increases at a constant rate of 1 for each increase of 1 of x. Any parabola will show this same constant change of slope although the actual values will vary.

Suppose in our original example about the bolts, that the graph of n against time had been a parabola instead of a straight line. What could we have said about the rate of use of bolts? The rate of use m would now be changing at a constant rate. Surely this is what we would call a constant *acceleration*?

6.3   The equation of the parabola

We have seen that a parabola is described by the equation

$$y = ax^2 + bx + c \tag{44}$$

If the graph of number of bolts against time had been parabolic in shape then the equation connecting these two variables would have taken the form

$$n = at^2 + bt + C, \tag{45}$$

(m no longer appears in the equation as it is no longer a constant).

Without using the graph, the value of n at any particular value of t can easily be found by substitution of the appropriate value of t in the equation. But suppose we wanted to find the value of t when n took a particular value. How would we solve the equation for t?

Rearranging equation (45) gives

$$at^2 + bt + (C - n) = 0,$$

or

$$at^2 + bt + K = 0. \tag{46}$$

44

Equation (46) is called a *quadratic* equation and in its general form it is usually written

$$ax^2 + bx + c = 0. \tag{47}$$

The problem now is to solve the equation for x (the general variable).

First, we divide both sides of equation (47) by a, giving

$$x^2 + \frac{b}{a}x + \frac{c}{a} = 0.$$

or

$$x^2 + \frac{b}{a}x = -\frac{c}{a}. \tag{48}$$

For an idea of what to do next, let us see which terms $(x + b/2a)^2$ gives when it is worked out. (You do it before going on.) It turns out that

$$\left(x + \frac{b}{2a}\right)^2 = x^2 + \frac{b}{a}x + \frac{b^2}{4a^2}. \tag{49}$$

Comparing equations (48) and (49) it is clear that

$$\left(x + \frac{b}{2a}\right)^2 = \frac{b^2}{4a^2} - \frac{c}{a} = \frac{b^2 - 4ac}{4a^2}.$$

Taking the square root of each side and remembering that a positive quantity has a positive and a negative square root, we get

$$x + \frac{b}{2a} = \pm\sqrt{\frac{b^2 - 4ac}{4a^2}},$$

or

$$x = \frac{-b \pm \sqrt{(b^2 - 4ac)}}{2a}. \tag{50}$$

We have now solved the equation for x in terms of the constants.

If $b^2 > 4ac$, the two roots are real and different.

If $b^2 = 4ac$ the roots are real and both equal to $-b/2a$.

If $b^2 < 4ac$ the expression in the square root sign is negative and since there is no real quantity whose square is negative, the roots in this case are said to be imaginary.

Equation (50) is quite general and can always be used to obtain the roots of a quadratic equation. If, however, *factors* of the left-hand side of equation (47) can be found, the roots are more easily obtained by setting each of the factors in turn equal to zero and solving the resulting simple equations. This process is illustrated in the exercise that follows.

Exercise

Solve the equations:

(i) $2x^2 + 5x - 12 = 0$;

(ii) $x^2 + 11 \qquad = 7x.$

(i) The left-hand side of this equation has factors $(2x - 3)$ and $(x + 4)$, so that the equation can be written

$$(2x - 3)(x + 4) = 0.$$

Hence either

$$(2x - 3) = 0$$

or

$$(x + 4) = 0,$$

giving

$$x = \tfrac{3}{2} \quad \text{or} \quad -4.$$

(ii) To obtain a solution using equation (51), the values $a = 1$, $b = -7$, $c = 11$ must be substituted, giving

$$x = \frac{7 \pm \sqrt{(7^2 - 4 \times 1 \times 11)}}{2 \times 1} = \frac{7 \pm \sqrt{5}}{2}$$

$$= \frac{7 \pm 2.236}{2}.$$

Thus

$$x = 4.618 \quad \text{or} \quad 2.382.$$

If you cannot remember the formula, there is an alternative way of calculating the values of x for the equation in (ii). It must first be rearranged to

$$x^2 - 7x = -11.$$

Each side can then be made a perfect square. We can write

$$(x - \tfrac{7}{2})^2 + \text{some number} = -11.$$

When the squared term is expanded we get $x^2 - 7x + \frac{49}{4}$, so the number we have to add to restore the original equation must be $-\frac{49}{4}$. Hence

$$(x - \tfrac{7}{2})^2 - \tfrac{49}{4} = -11,$$
$$(x - \tfrac{7}{2})^2 = -11 + \tfrac{49}{4} = \tfrac{5}{4}.$$

Thus

$$x - \frac{7}{2} = \frac{\pm \sqrt{5}}{2}$$

$$x = \frac{+\sqrt{5}}{2} + \frac{7}{2} \quad \text{or} \quad x = \frac{-\sqrt{5}}{2} + \frac{7}{2},$$
$$x = 4.618 \quad \text{or} \quad 2.382.$$

SAQ 14

(a) Find the value of k for which the equation $4x^2 - 8x + k = 0$ has equal roots.

(b) Solve the equations:
   (i) $8x^2 - 2x - 3 = 0$;
   (ii) $5x^2 + 10 = 17x$.

(c) Show that the equation $kx(1 - x) = 1$ has no real roots if $k < 4$.

6.4 Simultaneous equations

A number of separate equations are often needed to specify all of the relationships between several variables. For all the variables to be 'solved', the number of separate equations must equal the number of variables. For example, suppose two variables x and y simultaneously satisfy a pair of linear equations such as

$$3x + 4y = 7$$
$$2x - y = 1$$

in which both equations are of the first degree (there is no power of a variable greater than 1). By manipulating them we can discover what x and y actually are.

46

The second equation of the pair gives

$$y = 2x - 1.$$

This can now be substituted in the first giving

$$3x + 4(2x - 1) = 7$$

or

$$3x + 8x - 4 = 7.$$

Thus

$$11x = 11$$
$$x = 1,$$

hence

$$y = 1.$$

In this section it is assumed that you are already familiar with the solution of pairs of equations such as these. We shall only consider pairs of equations in which at least one is of a higher degree than the other and where the solution can be made to depend on solving a quadratic equation. Few fixed rules can be laid down but some of the methods available are illustrated in the following exercises.

Exercise A

Solve the simultaneous equations:

$$xy = 10,$$
$$3x + 2y = 16.$$

When one of the equations is of the first degree, either unknown is easily expressed in terms of the other. Substitution in the second equation then results in a single equation in one unknown.

The second equation of this pair gives

$$y = 8 - \frac{3x}{2}.$$

Substituting in the first equation gives

$$x\left(8 - \frac{3x}{2}\right) = 10$$

which, after multiplication by 2 and slight rearrangement, can be written

$$3x^2 - 16x + 20 = 0,$$

or

$$(3x - 10)(x - 2) = 0.$$

Thus

$$x = \tfrac{10}{3}$$

or

$$x = 2,$$

and since

$$y = 8 - \frac{3x}{2},$$

the corresponding values of y are 3 and 5.

Exercise B

Solve the pair of equations:

$$x^2 + 4xy + y^2 = 13,$$
$$2x^2 + 3xy = 18.$$

When the two equations are of the same degree in both y and x, and when the separate terms involving the unknowns are all of this degree, the solution can be obtained by first substituting $y = ax$.

The two equations in the pair given then become:

$$x^2(1 + 4a + a^2) = 13,$$
$$x^2(2 + 3a) = 8.$$

Dividing these two equations gives

$$\frac{1 + 4a + a^2}{2 + 3a} = \frac{13}{8},$$

and by cross-multiplication

$$8(1 + 4a + a^2) = 13(2 + 3a)$$
$$8a^2 - 7a - 18 = 0$$
$$(a - 2)(8a + 9) = 0.$$

Thus

$$a = 2 \quad \text{or} \quad -\tfrac{9}{8}.$$

The value of x can now be obtained by substitution in one of the original equations. Choosing the second, and substituting $a = 2$ gives

$$x^2(2 + 3 \times 2) = 8$$

or

$$x^2 = 1,$$

so that

$$x = \pm 1.$$

Since $y = ax$ and $a = 2$, the corresponding values of y are ± 2.

The second value $-\tfrac{9}{8}$ for a gives similarly

$$x^2[2 + 3(-\tfrac{9}{8})] = 8$$

leading to a negative value for x^2. There are thus no real solutions corresponding to this value of a.

SAQ 15

Solve the simultaneous equations:

(a) $\quad 2x - y = 5,$
$\qquad x^2 + xy = 2;$

(b) $\quad x^2 + y^2 = 5,$
$\qquad xy = 2.$

The solutions of simultaneous equations give the possible positions on a set of graphical axes where two (or more) curves cross or meet. The fact that two simultaneous *linear* equations have *one* solution for x and *one* for y simply means that the two straight line graphs described by the two equations cross at one point only, as shown in Figure 35.

If the pair of equations consist of one linear and one equation of the second degree (say a parabola) then there will probably be *two* points where the graphs cross giving *two* pairs of solutions, as shown in Figure 36.

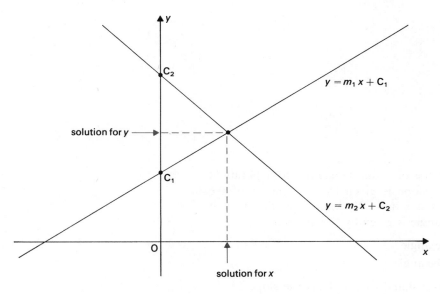

Figure 35   Solution of two linear simultaneous equations

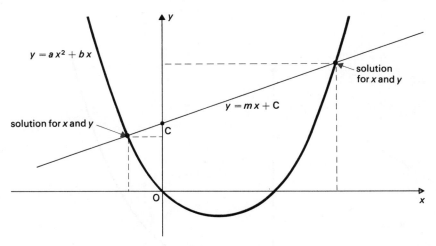

Figure 36   Solution of a pair of simultaneous equations where one is linear, one parabolic

For equations of a higher degree, the number of sets of solutions will be equal to the degree.

Where there are no real solution or solutions (that is the solutions are 'imaginary' values), it simply means that the two graphs do not meet at all. Two non-parallel straight lines will always meet, so the solution to linear simultaneous equations is always real. Figure 37 shows the situation for a pair of simultaneous equations where one is linear, one is parabolic (second degree) and the solutions are imaginary.

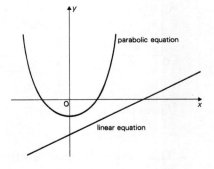

Figure 37

CALCULUS

7.1 Differentiation

When we were discussing the slope of a line tangential to a parabola in Section 6.2, we said that the slope is given by the tangent of the angle that the line makes with the x axis. We know that the tangent of an angle in a right-angled triangle is given by the quotient

$$\frac{\text{length of side opposite the angle}}{\text{length of side adjacent to the angle}},$$

so in Figure 38 the tangent of θ is equal to y/x. Thus the slope of the line OA is y/x.

In Figure 39, the slope of the line AB is given by

$$\tan \theta = \frac{\text{BC}}{\text{AC}} = \frac{y}{\text{AD}}.$$

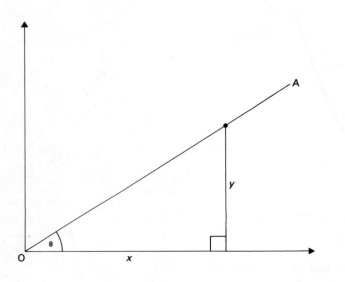

Figure 38

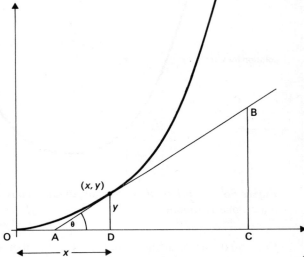

Figure 39

This section discusses a method of determining the slope which does not involve drawing.

Suppose that in Figure 40 the slope at the point $P(x, y)$ to the curve $y = x^2$ is required. This means that the slope of the tangent at this point must be found. The diagram shows two chords, PQ_1 and PQ_2.

The slope of PQ_2 is $(y_2 - y)/(x_2 - x)$ and is clearly nearer the slope of the tangent at P than is the slope of PQ_1, which equals $(y_1 - y)/(x_1 - x)$. If a third point Q_3 is taken on the curve, between P and Q_2, the slope of PQ_3 will be an even closer approximation to the slope of the tangent at P.

It appears that as the point Q moves down the curve towards P, so the slope of the chord PQ approaches the slope of the required tangent.

This process of moving Q down the curve towards P can be made to yield a result for the slope of the tangent, and hence the slope at P.

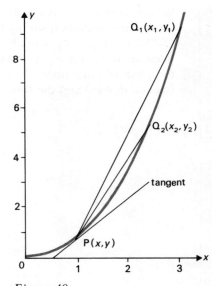

Figure 40

Look at Figure 41. Suppose Q is so close to P that its co-ordinates can be described as $(x+\delta x, y+\delta y)$, where δx (delta x) is a very small increase in x and δy is a very small increase in y. (Note that δx does not mean delta times x; it is one symbol, meaning a very small change in x.)

We have been considering the curve $y = x^2$, and as both P and Q lie on the curve, the co-ordinates of both P and Q satisfy the equation $y = x^2$.

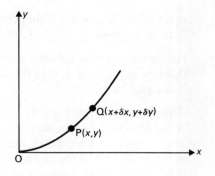

Since the co-ordinates of P are (x, y),

for   P: $y = x^2$.

Since the co-ordinates of Q are $(x+\delta x, y+\delta y)$,

Figure 41

for   Q: $y+\delta y = (x+\delta x)^2$
$$= x^2 + 2x\delta x + (\delta x)^2.$$

If $y+\delta y = x^2 + 2x\delta x + (\delta x)^2$

and also $y = x^2,$

then $\overline{\delta y = 2x\delta x + (\delta x)^2}$ (51)

by subtraction.

The slope of PQ is $\dfrac{(y+\delta y)-y}{(x+\delta x)-x} = \dfrac{\delta y}{\delta x}$

Dividing both sides of equation (51) by δx,

$$\frac{\delta y}{\delta x} = 2x + \delta x.$$ (52)

So the slope of PQ equals $2x + \delta x$.

As Q moves down the curve, nearer and nearer to P (thus making the slope of PQ more and more like that of the curve at P as shown in Figure 42), δx gradually gets very close to zero. We say that δx tends to zero, and this is written as

$\delta x \to 0$ (read 'delta x tends to zero').

It will also be true that $\delta y \to 0$, although not necessarily at the same rate.

Whether or not δx ever actually reaches zero is not important. What does matter in the case of functions whose y–x curves are continuous and smooth is that as δx approaches zero, the *ratio* $\delta y/\delta x$ gets closer and closer to some particular limiting value until finally it becomes equal to that value. This is a very special relationship in mathematics and forms the basis of calculus.

To summarize the idea we can write:

as $\delta x \to 0$, $\dfrac{\delta y}{\delta x} \to$ some limiting value.

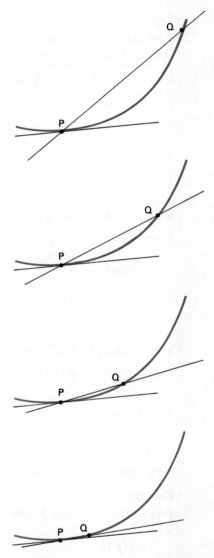

Figure 42   The slope of PQ gets nearer to the slope at P as Q approaches P

derivative, differential

We use a special notation to denote this limiting value: it is the term dy/dx (read 'dee-y by dee-x'), which is called the *derivative* or *differential* of y with respect to x. We can now write:

as $\delta x \to 0$, $\dfrac{\delta y}{\delta x} \to \dfrac{dy}{dx}.$

You have seen in the previous example that the left-hand side of equation (52) ($\delta y/\delta x$) became $2x$, provided $\delta x \to 0$. So if

$y = x^2$

then

$$\frac{dy}{dx} = 2x$$

51

everywhere, regardless of the actual value of y or x at which the differential is to be applied.

Hence we can now actually define the slope of a curve *at a point* using this new expression dy/dx: the slope of the curve $y = x^2$ at any *point* on it is equal to $2x$.

The process of obtaining dy/dx in this way is called *differentiation from first principles*. In the next section you will meet some rules which will enable you to find dy/dx much more quickly. This process is called *differentiation*. It is the process by which the slope of the curve (i.e. dy/dx) is found when the equation of the curve is known. The function is said to have been 'differentiated' to find dy/dx.

differentiation from first principles

differentiation

Exercise A

Find the slope of the curve $y = x^2$ at the points $(1, 1)$, $(-2, 4)$ and $(2.5, 6.25)$.

At the point $(1, 1)$, $x = 1$ and $y = 1$.

The slope of the curve is

$$\frac{dy}{dx} = 2x$$

$$= 2 \times 1$$

$$= 2.$$

At the point $(-2, 4)$, $x = -2$ and $y = 4$.

The slope of the curve is

$$\frac{dy}{dx} = 2x$$

$$= 2 \times (-2)$$

$$= -4.$$

At the point $(2.5, 6.25)$, $x = 2.5$ and $y = 6.25$.

The slope of the curve is

$$\frac{dy}{dx} = 2x$$

$$= 2 \times 2.5$$

$$= 5.$$

The exercise illustrates that $dy/dx = 2x$ is an equation which can be used to find the slope at any point on the curve $y = x^2$.

SAQ 16

SAQ 16

Figure 43 shows the curve $y = x^2$ with tangents drawn at $x = -2$, $x = 1$ and $x = 2.5$. Use the graph to find the values of the slope of the curve at these points. Show that they agree with the values obtained in Exercise A in this section by differentiation.

Exercise B

Find the slope of the curve (i.e. find dy/dx) for $y = x^3 + 4x - 2$.

In Figure 44, P is the point (x, y) and Q is the point $(x + \delta x, y + \delta y)$, where P and Q lie on the curve $y = x^3 + 4x - 2$. Since Q lies on the curve,

$$y + \delta y = (x + \delta x)^3 + 4(x + \delta x) - 2.$$

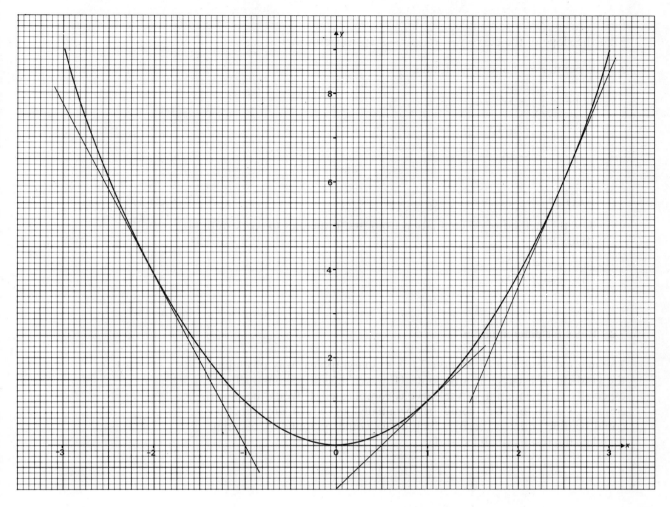

Figure 43

Multiplying out the brackets,

$$y + \delta y = x^3 + 3x^2\delta x + 3x(\delta x)^2 + (\delta x)^3 + 4x + 4\delta x - 2 \quad \text{(for Q)},$$
$$\underline{y = x^3 + 4x - 2 \quad \text{(for P)},}$$
$$\delta y = 3x^2\delta x + 3x(\delta x)^2 + (\delta x)^3 + 4\delta x \quad \text{(subtracting)}.$$

Dividing by δx,

$$\frac{\delta y}{\delta x} = 3x^2 + 3x\delta x + (\delta x)^2 + 4. \qquad (53)$$

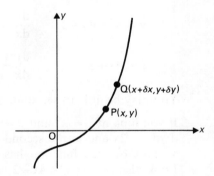

Figure 44

In the limit as Q approaches P and $\delta x \to 0$:

$$\underset{\delta x \to 0}{\text{limit}} \frac{\delta y}{\delta x} = \frac{\mathrm{d}y}{\mathrm{d}x} = \text{the slope of the curve at P};$$

$3x^2$ stays the same, as it does not depend on δx or δy;

$3x\delta x \to 0$;

$(\delta x)^2 \to 0$;

4 stays the same, as it does not depend on δx or δy.

Putting all this in equation (53), gives the slope of the curve

$$\frac{\mathrm{d}y}{\mathrm{d}x} = 3x^2 + 4.$$

This result can be used in the following exercise.

Exercise C

Find the slope of the curve $y = x^3 + 4x - 2$ at the point $(2, 14)$.

At this point, $x = 2$ and $y = 14$, thus

$$\text{slope} = \frac{dy}{dx} = 3x^2 + 4$$

$$= 3(2)^2 + 4$$

$$= 16.$$

SAQ 17

SAQ 17

Use the method of differentiation from first principles to work out dy/dx for:

(a) $y = 4x + 3$;
(b) $y = x^2 - 2x$.

7.2 Rules for differentiation

The method described in Section 7.1 is a non-graphical way of finding the gradient of a curve at any point. However, it is a tedious method, especially if the equation of the curve is complicated. It would be much simpler if a set of rules could be formulated by which any function could be differentiated (i.e. by which dy/dx could be found). Collecting together the results obtained so far:

$$y = x^2 \qquad\qquad \frac{dy}{dx} = 2x;$$

$$y = x^2 - 2x \qquad\qquad \frac{dy}{dx} = 2x - 2;$$

$$y = x^3 + 4x - 2 \qquad\qquad \frac{dy}{dx} = 3x^2 + 4;$$

$$y = 4x + 3 \qquad\qquad \frac{dy}{dx} = 4.$$

You may be able to see some pattern emerging from these.

If you look at $y = x^2$ and $y = x^2 - 2x$, you will see that in the first case $dy/dx = 2x$ and in the second $dy/dx = 2x - 2$. In fact, it looks as if the x^2 part of each function has been changed to $2x$ by differentiation. Look also at $y = x^3 + 4x - 2$ and $y = 4x + 3$. The $4x$ has been changed to 4 by differentiation in both cases.

One rule of differentiation is:

> if y is a sum or difference of terms, then dy/dx is the sum or difference of the derivatives of these individual terms.

derivative of a sum of terms

If you look at $y = x^3 + 4x - 2$ and $y = 4x + 3$, you will see that in the first case $dy/dx = 3x^2 + 4$ and in the second $dy/dx = 4$. It has already been stated that 4 is the derivative of $4x$ and it seems reasonable to assume that $3x^2$ is the derivative of x^3. It means that the derivative of both -2 and $+3$ must be 0. This leads to a second rule:

> if y is a constant (a number not involving x), then $dy/dx = 0$.

derivative of a constant

The third rule is:

> if $y = ax^n$, then $\dfrac{dy}{dx} = nax^{n-1}$.

derivative of $y = ax^n$

54

The third rule can be checked for $y = x^3$, because it has already been shown that this differentiates to $3x^2$. In this case, $a = 1$ and $n = 3$, so the rule says

$$\frac{dy}{dx} = 3 \times 1 \times x^{3-1}$$

$$= 3x^2,$$

which is the result expected.

Exercise A

Check the third rule for the other terms already met: x^2, $4x$ and $2x$.

For x^2, $a = 1$ and $n = 2$, so

$$\frac{dy}{dx} = 2 \times 1 \times x^{2-1}$$

$$= 2x,$$

which is the expected result.

For $4x$, $a = 4$ and $n = 1$, so

$$\frac{dy}{dx} = 1 \times 4 \times x^{1-1}$$

$$= 4 \quad \text{(since } x^0 = 1\text{).}$$

For $2x$, $a = 2$ and $n = 1$, so

$$\frac{dy}{dx} = 1 \times 2 \times x^{1-1}$$

$$= 2.$$

The rule 'if $y = ax^n$, then $dy/dx = nax^{n-1}$' can be extended to the cases where n equals a fraction, or a negative number or both.

Exercise B

Differentiate $y = 3x^{1/2} \quad (x \neq 0)$.

$$\frac{dy}{dx} = 3 \times \tfrac{1}{2} \times x^{-1/2} = \tfrac{3}{2}x^{-1/2} = \frac{3}{2\sqrt{x}}.$$

Exercise C

Differentiate $y = x^{-2} + 2x^{-4} \quad (x \neq 0)$.

$$\frac{dy}{dx} = -2x^{-3} - 8x^{-5}.$$

There are other rules for finding derivatives, some of which are set out in Table 1. All of these could be arrived at by a process of differentiating from first principles, though this is tedious and seldom done. Notice that it is just as possible to differentiate trigonometric, exponential and logarithmic functions, as it is to differentiate the simple algebraic functions already discussed.

SAQ 18 **SAQ 18**

1 Use Table 1 to work out dy/dx for:
   (a) $y = x^{3/2} + \sin x$;
   (b) $y = 2\cos 2x - 3\sin 3x$.

2   Use Table 1 to work out dy/dx for:
   (a) $y = 2\log_e x - 2x^4$ $(x > 0)$;
   (b) $y = x^{-2} - x^{-4} + 2x^{-6}$ $(x \neq 0)$;
   (c) $y = x^{5/2} - 2x^{-1/2}$ $(x \neq 0)$.

Table 1   Derivatives

y	$\dfrac{dy}{dx}$	Comment
C (a constant)	0	
ax^n	nax^{n-1}	
$\sin ax$	$a\cos ax$	
$\cos ax$	$-a\sin ax$	
$\tan x$	$\dfrac{1}{\cos^2 x}$	$x \neq \begin{cases} \pi/2, 3\pi/2, \ldots \\ -\pi/2, -3\pi/2, \ldots \end{cases}$
$\log_e x$	$\dfrac{1}{x}$	$x > 0$
e^x	e^x	refer to Appendix A

7.3   Distance, velocity and acceleration

So far the derivative dy/dx has been linked to the slope of a curve $y = f(x)$. In this section the use of differentiation is extended.

Figure 45 shows the total distance s travelled by a train from its starting point plotted against time t.

In this case, the slope will be referred to as ds/dt, rather than dy/dx, because s and t are the variables in this example: Figure 45 is a graph of how s varies with t rather than of how y varies with x.

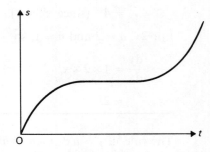

Figure 45   Distance–time graph

SAQ 19

Use the table of derivatives (Table 1) to calculate ds/dt for:
(a) $s = 2t^2 - t + 1$;
(b) $s = \cos t$.

Exercise A

Is the train in Figure 45 moving at a steady velocity? How can you judge?

No, it is not moving at a steady velocity. Its velocity is fairly steady at first, but it decreases to zero for a while before rising again to about its original value. You can see this because its total distance from its starting point does not increase at all at one stage, meaning it is stationary.

It is possible to judge the velocity by looking at the slope of the curve. Where the slope is steep the velocity is greater than it is where the slope is gentle.

The answer to this exercise will have shown you that it is possible to comment on the velocity of the train by looking at its distance–time graph.

It can be shown that at any point the value of ds/dt (the slope) is the value of the velocity v at that point.

$$v = \frac{ds}{dt} \tag{54}$$

56

Since ds/dt is of the form distance $\div$ time and the general formula for calculating velocity is

(distance travelled) $\div$ (time taken),

this linking of v to ds/dt should seem quite reasonable.

Exercise B

An object moves from rest so that its distance s from its starting point is given by the equation $s = t^2 + 2t$. Find its velocity:
(a) at $t = 1$;
(b) at $t = 2$;
(c) at $t = 3$.

From equation (54), $v = ds/dt$, so

$$v = \frac{ds}{dt} = 2t + 2.$$

At $t = 1$, $v = (2 \times 1) + 2 = 4$.
At $t = 2$, $v = (2 \times 2) + 2 = 6$.
At $t = 3$, $v = (2 \times 3) + 2 = 8$.

SAQ 20 SAQ 20

A car moves according to the equation $s = 4 - 2t$.
(a) Evaluate the velocity from the equation.
(b) Plot a graph of $s = 4 - 2t$ and thus try to explain the minus sign in your answer to (a).

The answer to SAQ 20 will have shown you that it is permissible to have a negative velocity in a problem. The minus sign is merely an indication of the direction of movement. In a similar way it is possible to have negative distances, or displacement. If displacements are called positive when they are in one direction from the datum point, they will be called negative when they are in exactly the opposite direction.

Acceleration is defined as the rate of change of velocity with time. So, using a as the symbol for acceleration,

$$a = \frac{dv}{dt}. \tag{55}$$

This formula is justifiable in the same way as is $v = ds/dt$. The equations $a = dv/dt$ and $v = ds/dt$ show that the acceleration is linked to the distance in some way. The equation for distance is differentiated once to obtain the velocity, and this equation for velocity is differentiated again to obtain the acceleration. This really means that the expression for s is differentiated twice to obtain the acceleration.

This is written as

$$a = \frac{d^2s}{dt^2} \quad \text{(read 'dee-two-s by dee-t-squared'),} \tag{56}$$

where d^2s/dt^2 is called the *second derivative* of s with respect to time.   **second derivative**

Exercise C

Find the acceleration if $s = 3t^3 + 2t - 1$.

$$v = \frac{ds}{dt} = 9t^2 + 2,$$

$$a = \frac{dv}{dt} = \frac{d^2s}{dt^2} = 18t.$$

A useful shorthand sometimes used is:

$$v = \frac{ds}{dt} = \dot{s}, \tag{57}$$

$$a = \frac{dv}{dt} = \dot{v}, \tag{58}$$

$$a = \frac{d^2s}{dt^2} = \ddot{s}. \tag{59}$$

The dot indicates differentiation with respect to time, and the number of dots indicates whether it is a first or second derivative. More dots would indicate a *higher derivative*, three dots meaning 'differentiated three times with respect to time' and so on.

SAQ 21

SAQ 21

(i) If $s = \sin t + \cos t$, find:
(a) $\dot{s}$ (b) $\ddot{s}$.
(ii) If $s = 3t^2 - 2t$, find the velocity and acceleration at time $t = 3$.

The units used for s, v, a and t in these equations must be consistent. For instance, if s is in metres and t is in seconds, then v is in metres per second and a is in metres per (second squared).

7.4 Further rules for differentiation

So far the functions differentiated have been fairly simple. Rules exist for dealing with more complicated functions. The first derivative of a function of x, $f(x)$ can be written as $f'(x)$. The second derivative as $f''(x)$.

1 *Product of functions*

If $y = f(x)g(x)$, where $f(x)$ and $g(x)$ are functions of x, then

$$\frac{dy}{dx} = f(x)g'(x) + f'(x)g(x). \tag{60}$$

In words:

the derivative of a product
= (the first) × (the derivative of the second)
+ (the derivative of the first) × (the second).

product of functions

Exercise A

Differentiate $y = x^3 \sin x$.

In this case

$$f(x) = x^3, \qquad \text{so } f'(x) = 3x^2;$$
$$g(x) = \sin x, \qquad \text{so } g'(x) = \cos x.$$

Thus from equation (60),

$$\frac{dy}{dx} = f(x)g'(x) + f'(x)g(x)$$

$$= x^3 \cos x + 3x^2 \sin x.$$

(Note the rearrangement of the last term – it is conventional and less confusing to put the power of x before the trigonometric function.)

2 Quotient of functions

If

$$y = \frac{f(x)}{g(x)}$$

where $f(x)$ and $g(x)$ are functions of x and $g(x)$ is not zero, then

$$\frac{dy}{dx} = \frac{f'(x)g(x) - f(x)g'(x)}{[g(x)]^2}. \qquad (61)$$

In words:

the derivative of a quotient
= [(the derivative of the numerator) × (the denominator)
− (the numerator) × (the derivative of the denominator)]
÷ (the denominator)^2.

Exercise B

Differentiate $y = \dfrac{4x^5}{6x+2}$.

$$f(x) = 4x^5, \qquad \text{so } f'(x) = 20x^4;$$
$$g(x) = 6x+2, \qquad \text{so } g'(x) = 6.$$

Thus

$$\frac{dy}{dx} = \frac{20x^4(6x+2) - 4x^5 \times 6}{(6x+2)^2}$$

$$= \frac{120x^5 + 40x^4 - 24x^5}{(6x+2)^2}$$

$$= \frac{96x^5 + 40x^4}{(6x+2)^2}.$$

3 Function of a function

This is an expression of the type $y = F[f(x)]$. Examples of functions of a function are:

$$(x^2+x)^4;$$
$$\sin(2x - \pi);$$
$$e^{x^2}.$$

The simplest way to explain how such functions are differentiated is by working through an exercise.

Exercise C

Differentiate $y = (x^2 + 1)^4$.

The first step is to put $w = x^2 + 1$. This makes $y = w^4$, so

$$\frac{dw}{dx} = 2x \quad \text{and} \quad \frac{dy}{dw} = 4w^3.$$

There is a rule of differentiation which states that

$$\frac{dy}{dx} = \frac{dy}{dw} \times \frac{dw}{dx}.$$

Using this, and substituting the values for dy/dw and dw/dx,

$$\frac{dy}{dx} = 4w^3 \times 2x.$$

Substituting back for w,

$$\frac{dy}{dx} = 4(x^2 + 1)^3 \times 2x$$

$$= 8x(x^2 + 1)^3.$$

The first step in differentiating a function of a function is always to put the second function equal to w (or any other letter) and then to use the rule

$$\frac{dy}{dx} = \frac{dy}{dw} \times \frac{dw}{dx}.$$

Exercise D

What would you put equal to w in order to differentiate:

(a) $y = \sin(2x - \pi)$;
(b) $y = e^{x^2}$;
(c) $y = (2t^{-1} - 3)^4$ $(t \neq 0)$;
(d) $y = \log_e t^2$ $(t \neq 0)$;
(e) $y = e^{\sin x}$.

(a) $2x - \pi$;
(b) x^2;
(c) $2t^{-1} - 3$;
(d) t^2;
(e) $\sin x$.

Exercise E

Differentiate $y = e^{\sin x}$

Your answer to the previous exercise has already shown you that $w = \sin x$.

$$w = \sin x, \quad \text{so} \quad y = e^w;$$

$$\frac{dw}{dx} = \cos x \quad \text{and} \quad \frac{dy}{dw} = e^w \quad \text{(see Table 1)};$$

thus

$$\frac{dy}{dx} = \frac{dy}{dw} \times \frac{dw}{dx}$$

$$= e^w \cos x.$$

Substituting for w gives

$$\frac{dy}{dx} = e^{\sin x} \cos x.$$

SAQ 22

1   Differentiate the following with respect to t or x as appropriate (using Table 1 to help you):

(a) $y = \sin x \cos x$;

(b) $y = \dfrac{x+2}{\cos x}$ $(\cos x \neq 0)$;

(c) $y = (2t^{-1} - 3)^3$ $(t \neq 0)$;

(d) $y = \dfrac{\sin t}{t^2 + 3}$.

7.5 Integration

In the section on differentiation it was shown that the velocity v and the acceleration a could be derived when an expression for the distance s

was known in terms of the time t. Sometimes there may be a problem where a is known in terms of t, and v and s are to be found.

It is often possible to find v and s from an equation for a by carrying out a process which can be thought of as the 'reverse' of differentiation.

Suppose

$$a = 4t.$$

Now

$$a = \frac{dv}{dt}$$

which means that an expression for v has been differentiated to give $4t$. If an expression can be found which differentiates to $4t$, then that expression should be the correct one for v. This process of finding v when dv/dt is known is called *integration*.

integration

Exercise A

Work out what v must have been in order that dv/dt should be $4t$. Remember that the exponent of t will have been reduced by 1 in the differentiation.

If $v = 2t^2$, then $dv/dt = 4t$.

This answer indicates that $v = 2t^2$ would differentiate to give $a = dv/dt = 4t$, and at first glance, it appears that the expression for v is now settled. However, look at Exercise B.

Exercise B

Differentiate the following:
(a) $v = 2t^2$;
(b) $v = 2t^2 + 1$;
(c) $v = 2t^2 + 2$;
(d) $v = 2t^2 + 200$;
(e) $v = 2t^2 - 60$.

For all the parts (a) to (e), $dv/dt = 4t$.

The answer shows that v found from $dv/dt = 4t$ can have a wide range of values. All the values differ from each other by a constant (a number), but not by terms involving t.

This arises because a constant always differentiates to zero. Therefore, instead of being able to find v completely when a is known, it is only possible to find it to within a constant.

If

$$a = 4t$$

then

$$v = 2t^2 + c,$$

where c is known as the *constant of integration*.

constant of integration

It is also possible to use integration to find s when ds/dt is known, y when dy/dx is known, and so on. However, the answer should always contain a constant of integration. Familiarity with the rules of differentiation means that simple integration can be readily carried out.

Exercise C

Find s if $v = 3t^2 - t + 3$.

Taking the expression for $v = \mathrm{d}s/\mathrm{d}t$ one term at a time:

$3t^2$ is the derivative of t^3;
$t = \frac{1}{2}(2t)$ and $2t$ is the derivative of t^2, so t is the derivative of $\frac{1}{2}t^2$;
3 is the derivative of $3t$.

Thus

$$s = t^3 - \tfrac{1}{2}t^2 + 3t + c. \tag{63}$$

The occurrence of c can be thought of in the following way. Equation (63) tells us how s, the distance of the object from a datum point, changes as time goes on, but it does not tell us where the object was when measurements were first taken (i.e. when $t = 0$).

Suppose distances are being measured from the datum point A (Figure 46) and that the object was 5 units of length from A when measurement started. Then when $t = 0$, $s = 5$. Putting these values into equation (63) gives

$$5 = 0 - 0 + 0 + c,$$

so

$$5 = c.$$

Therefore, under these initial conditions, $c = 5$.

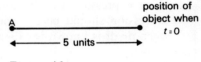

Figure 46

SAQ 23

SAQ 23

In equation (63), what would c be if the object was 3 units of distance from A when measurements started?

It is thus often possible to evaluate the constant of integration and hence find v or s fully by stating known conditions about the motion.

Exercise D

Find expressions for v and s if

$$a = 4t^2 + 3t,$$

and $v = 3$ and $s = 0$ when $t = 0$.

To find the expression for v:

$4t^2$ integrates to $\dfrac{4t^3}{3}$ and $3t$ integrates to $\dfrac{3t^2}{2}$,

so

$$v = \frac{4t^3}{3} + \frac{3t^2}{2} + c. \tag{64}$$

The question states that $v = 3$ when $t = 0$. Putting this in equation (64) gives

$$3 = 0 + 0 + c,$$

so $3 = c$ and

$$v = \frac{4t^3}{3} + \frac{3t^2}{2} + 3.$$

To find the expression for s:

$\dfrac{4t^3}{3}$ integrates to $\dfrac{t^4}{3}$, $\dfrac{3t^2}{2}$ integrates to $\dfrac{t^3}{2}$, and 3 integrates to $3t$;

so

$$s = \frac{t^4}{3} + \frac{t^3}{2} + 3t + k. \tag{65}$$

The question states that $s = 0$ when $t = 0$. Putting this in equation (65) gives

$$0 = 0 + 0 + 0 + k,$$

so

$$0 = k$$

and

$$s = \frac{t^4}{3} + \frac{t^3}{2} + 3t.$$

SAQ 24
SAQ 24

Consider $s(t)$, $v(t)$ and $a(t)$ to be distance, velocity and acceleration respectively, all functions of time t.

1   To within a constant of integration, find v for the following:
    (a) $a = 6t^2$;
    (b) $a = 3 + 2t$;
    (c) $a = 5 - 3t^3$.

2   To within a constant of integration, find s for the following:
    (a) $v = t^4 + t^3 + t^2 + t + 1$;
    (b) $v = 3t^2 - 10t + 6$.

3   Find v in the following:
    (a) $a = 2t^3 + t$, if $v = 2$ when $t = 0$;
    (b) $a = t - 2$, if $v = 0$ when $t = 1$.

4   Find v and s if
    $a = 4t^3 - 3t^2 + 2$, and $v = 0$ and $s = 6$ when $t = 0$.

There is a notation used for integration, which is as follows. Consider some function of time $F(t)$.

If

$$\frac{dv}{dt} = F(t),$$

then

$$v = \int F(t)\, dt$$

where $\int$ means 'the integral of' and dt indicates that the integration should be made with respect to t.

Similarly, if

$$\frac{ds}{dt} = F(t),$$

then

$$s = \int F(t)\, dt;$$

or if

$$\frac{dy}{dx} = F(x),$$

then

$$y = \int F(x)\, dx.$$

These integrals are all called *indefinite integrals*.

For example, if

$$\frac{dv}{dt} = 4t^2 + 3t,$$

then

$$v = \int (4t^2 + 3t)\,dt.$$

As you have already seen in Exercise D in this section, this means that

$$v = \frac{4t^3}{3} + \frac{3t^2}{2} + c.$$

So far integration has been carried out by working out for each term what must have been differentiated in order to yield that term. It is possible to use a table of integrals, similar to Table 1 for derivatives, as shown in Table 2.

Table 2 Integrals

$f(x)$	$\int f(x)\,dx$	Comment
ax^n	$\dfrac{a}{n+1}x^{n+1} + c$	
$\sin ax$	$-\dfrac{1}{a}\cos ax + c$	
$\cos ax$	$\dfrac{1}{a}\sin ax + c$	
e^{kx}	$\dfrac{1}{k}e^{kx} + c$	
$\dfrac{1}{x}$	$\log_e x + c$	$x > 0$

7.6 Areas

Suppose a velocity–time graph for the movement of a car has the shape shown in Figure 47. Is it possible to find the distance travelled by the car between the times $t = 1$ and $t = 3$ by using the graph?

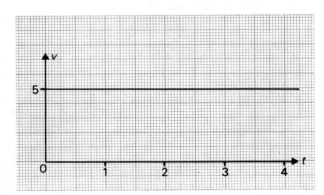

Figure 47   Velocity–time graph for $v = 5$

The last section has shown that the equation for the distance s can be found in terms of t by integration. We shall thus integrate first, and then see if the result can be related to the graph.

The velocity in this case is always equal to 5, so the equation of the graph is

$$v = 5.$$

Using the integral notation,

$$s = \int 5\,dt$$
$$= 5t + c,$$

where c is the constant of integration.

At time $t = 1$, the distance the car has travelled since starting is

$$5 \times 1 + c = 5 + c.$$

At time $t = 3$, the distance it has travelled since starting is

$$5 \times 3 + c = 15 + c.$$

So the distance it travels between $t = 1$ and $t = 3$ is

$$(15 + c) - (5 + c) = 15 + c - 5 - c$$
$$= 10.$$

Notice that c does not appear in this answer, since the distance travelled between $t = 1$ and $t = 3$ does not depend on the distance the car was from some datum point when $t = 0$, which is what c represents.

The calculation above is an example of the evaluation of a *definite integral*. The notation for the definite integral in this case is

definite integral

$$s = \int_1^3 5\,dt.$$

The numbers 1 and 3 on the integral sign show that the distance to be found is that between $t = 1$ and $t = 3$. The result of the integration is a number. This contrasts with an indefinite integral, where the result is a formula for s in terms of t.

In general, if $v = ds/dt = F(t)$ and the distance travelled s to be found is that between t_1 and t_2, then

$$s = \int_{t_1}^{t_2} F(t)\,dt, \tag{66}$$

where t_1 and t_2 are called the *limits* on the integral. The limit t_1 is the *lower limit* and t_2 is the *upper limit*.

limits,
lower limit, upper limit

To return to the original problem. Could the fact that the distance is 10 have been found from the graph?

Figure 48 is Figure 47 redrawn with the region under the line between $t = 1$ and $t = 3$ shaded. The area of this shaded region is

$$5 \times 2 = 10.$$

This is the value found for the distance by integration. Now

$$\text{velocity} = \frac{\text{distance travelled}}{\text{time taken}}$$

(provided the velocity is constant) and this can be rearranged to give

$$\text{distance travelled} = \text{velocity} \times \text{time taken}.$$

In the calculation of the area, 5 represents a velocity and 2 the time taken, so 10 does indeed represent the distance. Thus in this example, the area under the line gave the distance, and so did integration.

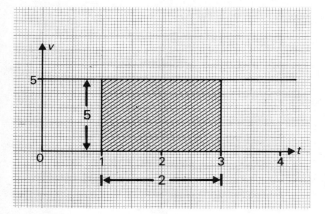

Figure 48

Take another example. Suppose that a car moves so that its velocity is given by the equation

$$v = 2t + 1.$$

Figure 49 shows a plot of this velocity–time relationship. What is the distance travelled between $t = 1$ and $t = 2.5$?

First, we can find it by integration:

$$s = \int_1^{2.5} (2t + 1)\, dt \quad \text{(using the definite integral notation)}$$
$$= [t^2 + t]_1^{2.5}.$$

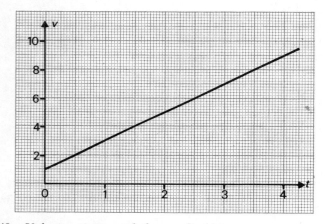

Figure 49   Velocity–time graph for $v = 2t + 1$

(By convention, the limits are indicated after the integration has been carried out, using the square bracket notation shown. Notice that, as before, c is not included: the distance travelled between $t = 1$ and $t = 2.5$ is independent of c.)

Substituting in the limits of time t,

$$s = [(2.5)^2 + 2.5] - [1^2 + 1]$$

(look back to the method used for finding the distance travelled in the last example, when $v = 5$, and you will understand what we are doing here).

So

$$s = [6.25 + 2.5] - [1 + 1]$$
$$= 6.75.$$

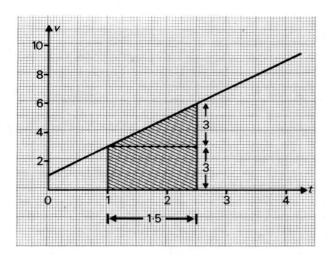

Figure 50

Now we shall find the distance by evaluating the area under the line. Figure 50 shows the required area shaded and split into a triangle and a rectangle.

$$\text{Area of triangle} = \tfrac{1}{2} \times \text{base} \times \text{height}$$
$$= \tfrac{1}{2} \times 1.5 \times 3$$
$$= 2.25.$$
$$\text{Area of rectangle} = 1.5 \times 3$$
$$= 4.5.$$
$$\text{Total area} = 6.75.$$

Once again, both integration and finding the area under the line yield the same value for the distance travelled.

This suggests that it may be possible to prove that the area under any velocity–time curve equals the distance, that the distance can also be found by integration, and so that the process of integration gives a value for the area under the curve.

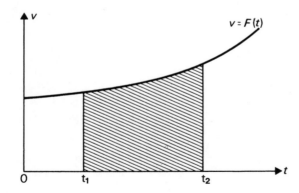

Figure 51   General velocity–time graph. The shaded area represents the distance and can be found by integration

In Figure 51 we would want to prove that the shaded area equals *s*, and that

$$s = \int_{t_1}^{t_2} F(t)\, \mathrm{d}t.$$

This can be done, but it is outside the scope of this unit. The proof indicates a much wider fact which is applicable to any curve, not just a

67

velocity–time curve: the area under any curve $y = f(x)$ between the x axis, x_1 and x_2 can be found by evaluating the integral

$$\int_{x_1}^{x_2} f(x)\,dx.$$

In Figure 52, the shaded area can be found by using the formula

$$\text{area} = \int_{x_1}^{x_2} f(x)\,dx.$$

In Figure 53, the shaded area can be found using the formula

$$\text{area} = \int_0^{t_1} G(t)\,dt$$

and so on.

However, it is not always possible to find the area under a curve in this way. It could not be done with the type shown in Figure 54 where the curve rises steeply, approaching infinity as x approaches zero. It is necessary to check that a given curve does not behave in this way in the region of interest before trying to evaluate the area.

Exercise A

An object moves so that it obeys the equation $v = \sin t$. Find the distance it moves:
(a) between $t = 0$ and $t = \pi$;
(b) between $t = \pi$ and $t = 2\pi$;
(c) between $t = 0$ and $t = 2\pi$.

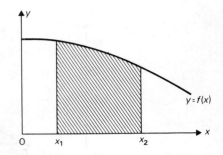

Figure 52

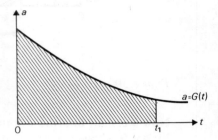

Figure 53

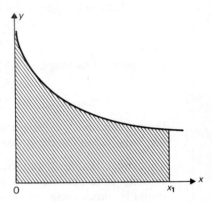

Figure 54

Using the result of equation (66):

for (a)

$$s = \int_0^{\pi} \sin t\,dt$$
$$= [-\cos t]_0^{\pi}$$
$$= -\cos \pi - [-\cos 0]$$
$$= -(-1) - (-1)$$
$$= 1 + 1$$
$$= 2;$$

for (b)

$$s = \int_{\pi}^{2\pi} \sin t\,dt$$
$$= [-\cos t]_{\pi}^{2\pi}$$
$$= -\cos 2\pi - [-\cos \pi]$$
$$= -1 - [-(-1)]$$
$$= -1 - 1$$
$$= -2;$$

for (c)

$$s = \int_0^{2\pi} \sin t\,dt$$
$$= [-\cos t]_0^{2\pi}$$
$$= -1 - (-1)$$
$$= -1 + 1$$
$$= 0.$$

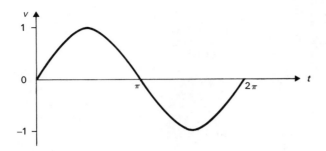

Figure 55   Graph of $v = \sin t$

To explain the results of this exercise a diagram of the motion $v = \sin t$ is useful, as shown in the graph of Figure 55.

Between $t = 0$ and $t = \pi$ the velocity is positive, so the object moves (say) from left to right, and the answer to part (a) shows it travels 2 units of distance in this time. Between $t = \pi$ and $t = 2\pi$ the velocity is negative, so the object is moving from right to left and the answer to part (b) shows that it travels back 2 units of distance during this time. Since it travels 2 units out and 2 back, its total displacement between times $t = 0$ and $t = 2\pi$ is 0, which is the result found in (c).

There are occasions, however, when the total distance travelled by an object, rather than the displacement, may be the required result. In the previous example the total distance travelled is 4 units between times 0 and 2π.

If the displacement of an object between times t_1 and t_2 is required, the expression for v is integrated between the limits t_1 and t_2.

If the total distance is required, the graph of the equation for v should be sketched. If it crosses the horizontal axis (say at t_0), the expressions for v should be integrated between t_1 and t_0, and again between t_0 and t_2. These two results are then added, ignoring the minus sign. If the curve does not cross the time axis, then the procedure is the same as that for finding the displacement.

Exercise B

Find both the displacement and the total distance travelled between $t = 0$ and $t = 2$ by an object moving according to the formula $v = t^3 - 1$.

Figure 56 shows a sketch of $v = t^3 - 1$. The curve crosses the t axis at $t = 1$.

$$\text{Displacement} = \int_0^2 (t^3 - 1)\,dt$$

$$= \left[\frac{t^4}{4} - t\right]_0^2$$

$$= [\tfrac{16}{4} - 2] - 0$$

$$= 2.$$

Distance between $t = 0$ and $t = 1$

$$= \int_0^1 (t^3 - 1)\,dt$$

$$= \left[\frac{t^4}{4} - t\right]_0^1$$

$$= [\tfrac{1}{4} - 1] - 0$$

$$= -\tfrac{3}{4}.$$

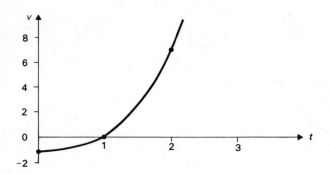

Figure 56   Graph of $v = t^3 - 1$

Distance between $t = 1$ and $t = 2$

$$= \int_1^2 (t^3 - 1) \, dt$$

$$= \left[\frac{t^4}{4} - t \right]_1^2$$

$$= \left[\frac{16}{4} - 2 \right] - \left[\frac{1}{4} - 1 \right]$$

$$= 2 - \left(-\frac{3}{4} \right)$$

$$= 2\frac{3}{4}.$$

Thus the total distance travelled $= \frac{3}{4} + 2\frac{3}{4} = 3\frac{1}{2}$.

SAQ 25   **SAQ 25**

Find the area between the curve $y = x^2$, the x axis, $x = 0$, and $x = 4$.

THE EXPONENTIAL FUNCTION

There are many examples in nature where the rate of increase of a number of items is proportional to the number already there. The numbers of most living creatures would increase in this way, were it not for such factors as predators and shortage of food. It is easy to visualize the shape of the graph of such a growth: it becomes progressively steeper.

Figure 57 shows a sketch of the shape such a curve should have. (Very roughly, the graph of the population of the world against time gives this shape of curve.)

In mathematics it is necessary to know more than just the shape of the curve: it is necessary to know the equation. Two curves which have approximately the right shape are

$$y = 2.5^t \quad \text{and} \quad y = 3^t.$$

where t is the time and y the number of items present. Figure 58 shows the graph of the first of these functions, Figure 59 shows the second.

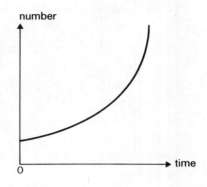

Figure 57

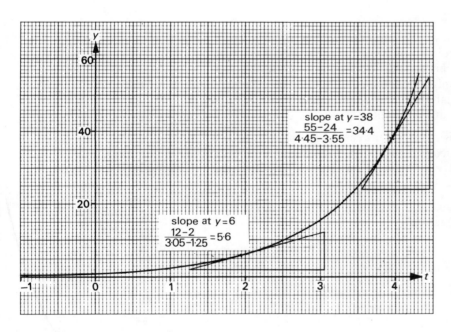

Figure 58   Curve $y = 2.5^t$

If the curve is to be such that the rate of increase is proportional to the number y already there, then

(rate of increase of y) $\propto y$

or

(rate of increase of y) $= ky,$ \hfill (66)

where k is the constant of proportionality. In the simplest case, $k = 1$ and

(rate of increase of y) $= y.$

so the curve must have its slope equal to y at each point. Such a curve is called the *exponential curve*.

exponential curve

71

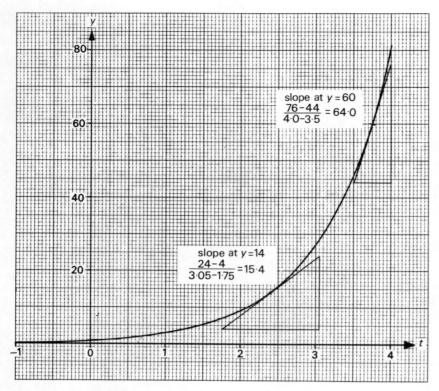

Figure 59   Curve $y = 3^t$

Figure 58 shows that the slope of the curve $y = 2.5^t$ is slightly lower than the y value at the two points considered. Figure 59 shows that the slope of $y = 3^t$ is slightly higher than the value of y at the two points. This suggests that there might be some number between 2.5 and 3, which can be called e, such that the slope of the curve $y = e^t$ at any point is equal to the value of y at that point.

In fact, such a number does exist. It is a non-recurring, non-terminating decimal whose value (to five decimal places) is 2.718 28. This number is called the *exponential* e, and $y = e^t$ always has a slope equal to the value of y. Figure 60 shows the curve $y = e^t$.

exponential e

In the general case, if k is not equal to 1 in equation (66) the exponential curve obtained will have the equation

$$y = e^{kt}.$$

The constant k affects the curve in the following way:

    if $k > 1$, the curve rises more steeply than Figure 60;
    if $k = 1$, the curve is that shown in Figure 60;
    if $k < 1$, the curve rises less steeply than Figure 60.

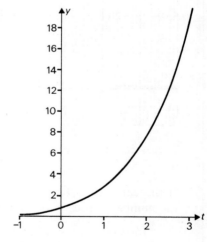

Figure 60   Curve $y = e^t$

A1   Differentiation and integration of e^x

If $y = e^x$, then $x = \log_e y$ and Table 1 gives

$$\frac{dx}{dy} = \frac{1}{y} = \frac{1}{e^x},$$

$$\frac{dy}{dx} = e^x = y.$$

Thus we have the important result that the differential of e^x is equal to e^x, that is

$$\frac{d}{dx}(e^x) = e^x. \tag{67}$$

This shows that the slope of the curve $y = e^x$ at any point is equal to the value of y at that point.

The integral of e^x,

$$\int e^x \, dx = e^x + c.$$

(68)

ANGLE IN A SEMICIRCLE

Considering triangle AOB in Figure 61,

> AO = BO

since both are radii of the same circle. Therefore

> angle OBA = angle OAB.

Now since

> angle BOZ + angle BOA = 180°

and the sum of all the angles in a triangle is 180°,

> angle BOZ = angle OBA + angle OAB

or

> angle BOZ = 2 × angle OAB (or OBA).

Similarly

> angle COZ = 2 × angle OAC (or OCA).

By addition

> angle BOZ + angle COZ = 2(angle OAB + angle OAC),

that is

$$180° = 2 \times \text{angle BAC}$$

or

> angle BAC = 90°.

Hence, the angle subtended by a diameter in a circle is equal to a right angle.

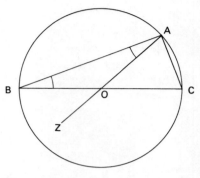

Figure 61

THE THEOREM OF PYTHAGORAS

The two big squares in Figures 62(a) and (b) are of exactly the same size. Inside each square four right-angled triangles have been arranged, each triangle with perpendicular sides of lengths a and b, and a hypotenuse of length c. In each case the remaining area of the big square is shaded.

The shaded part of the diagram (a) makes up two squares, one of side a and the other of side b. The shaded figure in diagram (b) has sides of length c. It too is a square, because its appearance is unchanged when the diagram is turned through a right angle. The shaded areas of the two diagrams must be equal, so

$$c^2 = a^2 + b^2.$$

This is the theorem of Pythagoras. It is often stated in words as 'The square on the hypotenuse of a right-angled triangle is equal to the sum of the squares on the other two sides'.

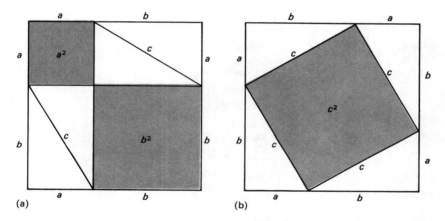

Figure 62   *Two arrangements of four identical triangles within a square*

ANSWERS TO SELF-ASSESSMENT QUESTIONS

SAQ 1

(a) $72AB + 18AC$.

(b) $A(A - B) + B(A - B)$
$= A^2 - AB + AB - B^2$
$= A^2 - B^2$.

(c) $3A[B + 2CD + BC] = 3AB + 6ACD + 3ABC$.

(d) $A - B\left[\dfrac{C}{D} + CE - \dfrac{CA^2}{C^2}\right] = A - \dfrac{BC}{D} - BCE + \dfrac{BA^2}{C}$.

SAQ 2

(a) (i) $\dfrac{2yz^4}{7x^3}$, (ii) $\dfrac{1}{y \times y^{4/3}} = \dfrac{1}{y^{7/3}}$.

(b) (i) $\left[\frac{81}{16}\right]^{3/4} = \sqrt[4]{\left(\frac{81}{16}\right)^3} = \left[\left(\frac{81}{16}\right)^{1/4}\right]^3 = \left(\frac{3}{2}\right)^3$
$= \frac{27}{8} = 3.375$.

(ii) $(64)^{-3/2} = \dfrac{1}{[(64)^{1/2}]^3} = \dfrac{1}{8^3}$
$= \frac{1}{512} = 0.00195$.

(c) $\dfrac{x^8 y^2 z^{-6} x^{-5} y^2 z}{x^{1/2} z^{1/2}} = \dfrac{x^3 y^4 z^{-5}}{x^{1/2} z^{1/2}} = \dfrac{x^{5/2} y^4}{z^{11/2}}$.

SAQ 3

(a) $\bar{3}.79934$, (b) 2.73878, (c) 0.33646,
(d) 1.63949, (e) $\bar{2}.25527$, (f) $\bar{4}.43136$.

SAQ 4

(a) 154.8, (b) 4.319, (c) 0.1967,
(d) 0.0008748, (e) 31.33, (f) 0.1294.

SAQ 5

(a) 0.07993, (b) 0.7504, (c) 0.1233.

SAQ 6

(a) 0.0031, (b) 0.03458, (c) 76.92.

SAQ 7

(a) 0.01615, (b) 0.2348, (c) 0.06040,
(d) 0.005804, (e) 50.69, (f) 24.40,
(g) 0.1599, (h) 0.0001129.

SAQ 8

(a) $\cos\theta$, $-\sin\theta$, $-\cot\theta$; (b) $\cos\theta$, $\sin\theta$, $\cot\theta$;
(c) $-\cos\theta$, $\sin\theta$, $-\cot\theta$; (d) $-\cos\theta$, $-\sin\theta$, $\cot\theta$.

SAQ 9

$$\tan(A + B) = \frac{\sin(A + B)}{\cos(A + B)}$$

$$= \frac{\sin A \cos B + \cos A \sin B}{\cos A \cos B - \sin A \sin B}.$$

Dividing numerator and denominator by $\cos A \cos B$ gives

$$\tan(A + B) = \frac{\tan A + \tan B}{1 - \tan A \tan B}.$$

Writing $-B$ in place of B gives

$$\tan(A - B) = \frac{\tan A + \tan(-B)}{1 - \tan A \tan(-B)}$$

$$= \frac{\tan A - \tan B}{1 + \tan A \tan B}.$$

SAQ 10

(a) Writing $B = A$ in the expression for $\tan(A + B)$ we get

$$\tan 2A = \frac{\tan A + \tan A}{1 - \tan A \tan A}$$

$$= \frac{2\tan A}{1 - \tan^2 A}.$$

(b) By the Theorem of Pythagoras

$$c^2 = a^2 + b^2.$$

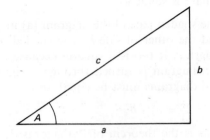

Figure 63

Since

$$\sin^2 A = \frac{b^2}{c^2}$$

and

$$\cos^2 A = \frac{a^2}{c^2},$$

$$\sin^2 A + \cos^2 A = \frac{b^2}{c^2} + \frac{a^2}{c^2} = \frac{a^2 + b^2}{c^2}.$$

But

$$a^2 + b^2 = c^2.$$

Thus

$$\sin^2 A + \cos^2 A = 1.$$

SAQ 11

(a) If $\sin\theta = 0.8660$, that is $\sqrt{3}/2$, $\theta = 60°$. Thus $\cos 60° = 0.5000$, that is $\frac{1}{2}$, and $\tan 60° = 1.7321$, that is $\sqrt{3}$.

(b) If $\cos\theta = -0.5000$, θ must be in the second and third quadrants (Figure 64). Since the angle whose cosine is 0.5000 (for $0° < \theta < 90°$) is $60°$, the angles whose cosine is -0.5000 must be $(180° - 60°) = 120°$ and $(180° + 60°) = 240°$.

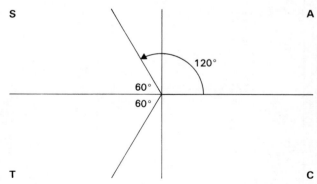

Figure 64

Note: there is another piece of shorthand that is used to denote the phrase in the above answer 'the angle whose cosine is'. It is the expression $\cos^{-1}$. This does not mean $1/\cos$. If a statement reads 'the angle whose cosine is 0.5000 is $60°$, it can be written as $60° = \cos^{-1} 0.5000$. Similarly for the other trigonometric functions sine, tangent, etc. we have $\sin^{-1}$, $\tan^{-1}$, etc.

(c) $\sin 330° = -\sin 30° = -\frac{1}{2}$;
$\cos 390° = +\cos 30° = \sqrt{3}/2$;
$\cos 570° = -\cos 30° = -\sqrt{3}/2$;
$\sin 510° = +\sin 30° = \frac{1}{2}$.

Therefore

$\sin 330° \cos 390° - \cos 570° \sin 510°$

$= -\sin 30° \cos 30° + \cos 30° \sin 30°$

$= 0$.

SAQ 12

Figure 65

For infinite value of C the graphs are somewhat difficult to draw!

SAQ 13

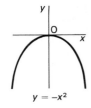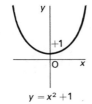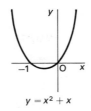

Figure 66

SAQ 14

(a) Here $a = 4$, $b = -8$, $c = k$. To give $b^2 = 4ac$, we must have

$(-8)^2 = 4 \times 4 \times k$

$k = \frac{64}{16} = 4$.

(b) (i) $x = \frac{3}{4}$ or $-\frac{1}{2}$;

(ii) $x = 2.643$ or 0.757.

(c) $kx(1-x) = 1$,

$kx - kx^2 - 1 = 0$,

$kx^2 - kx + 1 = 0$,

Thus $a = k$, $b = -k$, $c = +1$; and for $b^2 < 4ac$ (no real roots), we must have

$(-k)^2 < 4 \times k \times 1$

$k^2 < 4k$

$k < 4$.

SAQ 15

(a) $2x - y = 5$, (i)

$x^2 + xy = 2$. (ii)

From (i) $x = \dfrac{y+5}{2}$, $y = 2x - 5$.

Substituting in (ii) gives

$x^2 + x(2x-5) - 2 = 0$

$3x^2 - 5x - 2 = 0$

$(3x+1)(x-2) = 0$

$x = -\frac{1}{3}$ or $+2$.

Hence as $y = 2x - 5$,

$y = \dfrac{-17}{3}$ or -1

respectively.

(b) $x^2 + y^2 = 5$, (i)

$xy = 2$. (ii)

From (ii) $x = z/y$, and substituting in (i) gives

$\dfrac{4}{y^2} + y^2 = 5$,

$4 + y^4 - 5y^2 = 0$.

Let $Y = y^2$, then

$Y^2 + 5Y + 4 = 0$

$(Y-4)(Y-1) = 0$

$Y = 4$ or 1.

Thus

$y = \pm 2$ or ± 1.

Hence

$x = \pm 1$ or ± 2.

respectively.

SAQ 16

Figure 67 shows the slopes of the tangents calculated. You may have chosen different points on your tangents, but you should still have obtained answers that agree with those calculated in the exercise.

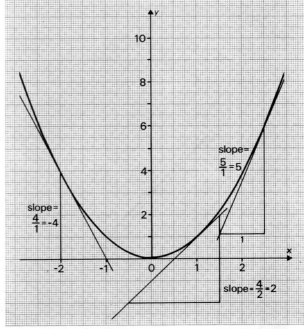

Figure 67   Graphs of $y = mx + C$

SAQ 17

(a) $y + \delta y = 4(x + \delta x) + 3$

$\delta y = 4\delta x$

77

$$\frac{\delta y}{\delta x} = 4.$$

$$\lim_{\delta x \to 0} \frac{\delta y}{\delta x} = \frac{dy}{dx} = 4.$$

(b) $\quad y + \delta y = (x + \delta x)^2 - 2(x + \delta x)$

$$\delta y = \delta x^2 + 2x\delta x - 2\delta x$$

$$\frac{\delta y}{\delta x} = \delta x + 2x - 2.$$

$$\lim_{\delta x \to 0} \frac{\delta y}{\delta x} = \frac{dy}{dx} = 2x - 2.$$

SAQ 18

1 (a) $\quad \dfrac{dy}{dx} = \dfrac{3x^{1/2}}{2} + \cos x;$

(b) $\quad \dfrac{dy}{dx} = -4 \sin 2x - 9 \cos 3x.$

2 (a) $\quad \dfrac{dy}{dx} = \dfrac{2}{x} - 8x^3;$

(b) $\quad \dfrac{dy}{dx} = -2x^{-3} + 4x^{-5} - 12x^{-7};$

(c) $\quad \dfrac{dy}{dx} = \frac{5}{2}x^{3/2} + x^{-3/2}.$

SAQ 19

(a) $\quad \dfrac{ds}{dt} = 4t - 1;$

(b) $\quad \dfrac{ds}{dt} = -\sin t.$

SAQ 20

(a) $\qquad s = 4 - 2t,$

so $\quad \dfrac{ds}{dt} = -2 \quad$ and $\quad v = -2.$

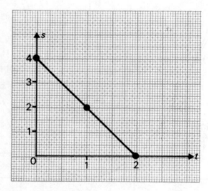

Figure 68

The distance of the object from the point from which measurements are being taken – the datum point – is decreasing, so at the start of the motion the object is moving back towards the point. Had it been moving out from it, the velocity would have been positive. As it is moving back, the velocity is negative.

SAQ 21

(i) (a) $\quad \dot{s} = \cos t - \sin t;$

(b) $\quad \ddot{s} = -\sin t - \cos t.$

(ii) $\qquad v = 16, \, a = 6.$

SAQ 22

(a) $\quad \dfrac{dy}{dx} = \cos^2 x - \sin^2 x;$

(b) $\quad \dfrac{dy}{dx} = \dfrac{\cos x + (x+2)\sin x}{\cos^2 x};$

(c) $\quad \dfrac{dy}{dx} = -6t^{-2}(2t^{-1} - 3)^2;$

(d) $\quad \dfrac{dy}{dx} = \dfrac{(t^2 + 3)\cos t - 2t \sin t}{(t^2 + 3)^2}.$

SAQ 23

$c = 3.$

SAQ 24

1 (a) $\quad v = 2t^3 + C;$

(b) $\quad v = 3t + t^2 + C;$

(c) $\quad v = 5t - \frac{3}{4}t^4 + C.$

2 (a) $\quad s = \frac{1}{5}t^5 + \frac{1}{4}t^4 + \frac{1}{3}t^3 + \frac{1}{2}t^2 + t + C;$

(b) $\quad s = t^3 - 5t^2 + 6t + C.$

3 (a) $\quad v = \frac{1}{4}t^4 + \frac{1}{2}t^2 + 2;$

(b) $\quad v = \frac{1}{2}t^2 - 2t + \frac{3}{2}.$

4 $\quad v = t^4 - t^3 + 2t;$

$\quad s = \frac{1}{5}t^5 - \frac{1}{4}t^4 + t^2 + 6.$

SAQ 25

$$\int_{x=0}^{x=4} x^2 \, dx = \left[\frac{x^3}{3}\right]_0^4 = \frac{64}{3} = 21\frac{1}{3}.$$

ACKNOWLEDGEMENTS

Grateful acknowledgement is made to the following for illustrations used in this unit:

Figures 2, 3 and 6 Illustrated London News; *Figure 4* British Railways Board; *Figure 5* from Hopkins, H. J. (1970) *A Span of Bridges*, David and Charles; *Figure 7* The Mansell Collection; *Figure 8* Camera Press, photo: Basil Williams; *Figure 9* from Thor Heyerdal (1958) *Aku-Aku* Allen and Unwin; *Figure 10* from Diderot, D. (1959) *L'Encyclopédie, ou Dictionnaire Raisonné des Sciences, des Arts et des Métiers* (English title: *A Diderot pictorial encyclopaedia of trades and industry*) Gillespie, C. G. (ed.), Dover; *Figure 11* G. Cussons Ltd; *Figure 12* Spirax Sarco Ltd.

Introduction to Engineering Mechanics